Risk Management of Strategic Readiness

Concepts, Practices, Data Analytics, and Next Steps

DWIGHT PHILLIPS, MINA POLLMANN, ANN MARIE DAILEY, PETER SCHIRMER, RYAN HABERMAN, BRETT ZAKHEIM, GWEN MAZZOTTA

Prepared for the Office of the Assistant Secretary of Defense for Readiness
Approved for public release; distribution is unlimited.

For more information on this publication, visit **www.rand.org/t/RRA3078-1**.

Published by the RAND Corporation, Santa Monica, Calif.
© 2025 RAND Corporation
RAND® is a registered trademark.

Library of Congress Cataloging-in-Publication Data is available for this publication.
ISBN: 978-1-9774-1473-1

Cover: Composite design by Carol Ponce adapted from images by Kylee Warren/U.S. Air National Guard and iamchamp/Adobe Stock.

About This Report

In 2023, the U.S. Department of Defense (DoD) codified the concept of strategic readiness to evaluate readiness through a lens that considered building capability for future conflicts while still meeting current missions.[1] DoD defines *strategic readiness* as "the ability to build, maintain, and balance warfighting capabilities and competitive advantages that ensure the DoD can achieve strategic objectives across threats and time horizons."[2] The DoD purpose for establishing the strategic readiness framework was to "inform senior leaders of the readiness trade-offs and impacts resulting from their decisions to better illuminate associated risks to force, mission, and the NDS [National Defense Strategy]."[3] One year into implementation, RAND researchers were asked to develop ways to ensure that department decision-making regarding strategic readiness is pedagogically coherent, objective, informed by data, and based on risk. Building on past DoD and RAND efforts, the authors of this report seek to advance the DoD purpose by summarizing key shortfalls in the application of the strategic readiness framework in DoD processes. This report provides recommendations for conceptualizing strategic readiness inputs, outputs, and outcomes; adopting risk management best practices for strategic readiness; and leveraging data science and artificial intelligence to better understand strategic readiness levels and forecast future ones.

The research reported here was completed in December 2024 and underwent security review with the sponsor and the Defense Office of Prepublication and Security Review before public release.

RAND National Security Research Division

This research was sponsored by the Office of the Assistant Secretary of Defense for Readiness and conducted within the Personnel, Readiness, and Health Program of the RAND National Security Research Division (NSRD), which operates the National Defense Research Institute (NDRI), a federally funded research and development center sponsored by the Office of the Secretary of Defense, the Joint Staff, the Unified Combatant Commands, the Navy, the Marine Corps, the defense agencies, and the defense intelligence enterprise.

For more information on the RAND Personnel, Readiness, and Health Program, see https://www.rand.org/nsrd/prh or contact the director (contact information provided on the webpage).

Acknowledgments

We would like to thank Assistant Secretary of Defense for Readiness Peter Belk for sponsoring this research and providing insights and comments throughout the research. We are grateful for the time and knowledge shared by the offices of the Deputy Assistant Secretaries of Defense for Force Readiness, Force Education and Training, and Force Safety and Occupational Health. In particular, we are indebted to Jud Crane, Lt Col David "Maddog" Galloway, and Ryan Crotty.

We would also like to thank Jim Mitre for his counsel and advice throughout the research. We also thank Mike Linick and the other reviewers of this report for their detailed feedback and excellent suggestions. RAND researchers Katharina Best and Ashley Rhoades provided helpful feedback on the later stages of our work. We would also like to thank the director and associate director of the RAND Personnel, Readiness, and Health Program, Molly McIntosh and Daniel Ginsberg, for their dedicated support and oversight of our study. Finally, we thank Melissa Bauman for her help in creating the graphics within this report.

Summary

One of the primary management functions of the U.S. Department of Defense (DoD) is managing Total Force preparedness to prevail in conflict and competition across different time horizons and strategic objectives. In 2023, DoD codified the concept of strategic readiness to evaluate readiness through a lens that considered building capability for future conflicts while still meeting current missions.[1] The new policy established a framework for assessing strategic readiness and assigned oversight functions. After a year of initial implementation, it is worth examining progress toward these goals and consider next steps.

Approach

RAND was asked to develop ways to ensure that department decisionmaking regarding strategic readiness is pedagogically coherent, objectively grounded, informed by data, and based on risk. Our report is structured around three questions:

- Is the strategic readiness concept sufficiently clear to be useful to DoD stakeholders?
- How can DoD risk management of strategic readiness be improved?
- How can DoD leverage ongoing data efforts to better inform its decisionmaking?

We approached these questions by reviewing DoD's policies, strategic guidance documents, and risk and readiness assessments. We conducted a literature review of RAND, academic, professional military, and business literature for best practices for military readiness, risk management, and data analytic methods. To gain first-hand insights and critiques, we conducted semi-structured interviews with key Office of the Secretary of Defense (OSD) and Joint Staff stakeholders, as well as a red team panel with subject-matter experts. Finally, we examined ongoing department data modeling efforts and experimented with using artificial intelligence (AI) to support analysis of strategic readiness.

Key Findings

Building on past DoD and RAND efforts, this report identified several key shortfalls in the strategic readiness framework's application in DoD processes:

---

[1] Department of Defense Instruction 3000.18, *Strategic Readiness*, U.S. Department of Defense, November 30, 2023.

- The DoD definition of strategic readiness needs to include an explanation of how its dimensions should be measured and how they interact as inputs and outputs leading to outcomes.
- DoD risk management of Total Force strategic readiness is inconsistent in its analytic methods, application in DoD processes, and DoD-wide oversight.
- Middle-term strategic readiness risks lack a champion in policy deliberations, both in internal DoD discussions and in national-level policy conversations.
- DoD has an abundance of data and readiness reporting but is challenged to turn those data and reporting into timely, issue-relevant assessments for policymakers because of uncertainties about data quality and a lack of data analytics tools.

These shortfalls—conceptual, risk management practices, and data analytics—challenge OSD's ability to create decision space for DoD senior leaders to confront, address, and mitigate strategic readiness gaps.

Recommendations

Using a review of best practices for concepts, risk management, and data analytics, we recommend that DoD take the following steps:

- Adopt a logic model for strategic readiness that maps inputs to outputs and then to outcomes to guide deliberations about policy trade-offs.
- Create an independent oversight function within DoD for strategic readiness.
- Develop clear guidance for risk thresholds, organizational responsibilities for risk management, and application of strategic readiness assessments in DoD processes.
- Develop the organizational expertise and analytic tools in OSD to dynamically track, assess, and explore strategic readiness trends within the DoD data ecosystem.
- Expand the suite of DoD data science tools—dashboards, modeling and simulation, and AI co-pilots—to enable risk analysts to leverage all relevant data ecosystems.

Next Steps

In this report, we have answered questions about what risk management of strategic readiness means and how to do it well. But in doing so, we have raised another question: Is DoD appropriately structured to implement these recommendations for risk management of strategic readiness? Although our research scope did not allow a fulsome examination of OSD organizational structure, we recommend that OSD explore several potential options:

- Place more emphasis on strategic readiness within the Office of the Under Secretary of Defense for Personnel and Readiness.

- Spin off oversight of strategic readiness to the Office of the Under Secretary of Defense for Policy.
- Create an Office of the Under Secretary of Defense for Strategic Readiness.
- Create an Office of the Assistant to the Secretary of Defense for Strategic Readiness that reports directly to the Deputy Secretary of Defense.

We did not conduct a thorough review of the pros and cons of each reorganization, and we lack sufficient analysis to make a recommendation. But our research suggests that internal reform of DoD organizational structures is likely necessary for improved risk management of strategic readiness and overall department strategic discipline.

Contents

Figures and Table

Figures

Table

CHAPTER 1

Introduction and Method

One of the primary management functions of the U.S. Department of Defense (DoD) is managing Total Force preparedness to prevail in conflict and competition across different time horizons and strategic objectives. In practical terms, DoD must manage trade-offs between consumption of warfighting capabilities today with building and preserving warfighting capabilities for the future, as well as make choices about what strategic objectives to prepare for.

In 2023, DoD codified the concept of strategic readiness to evaluate readiness through a lens that considered building capability for future conflicts while still meeting current missions.[1] In this report, we build on past DoD and RAND efforts to advance this DoD purpose by summarizing key shortfalls in the strategic readiness framework's application in DoD processes. This report provides recommendations for conceptualizing strategic readiness inputs, outputs, and outcomes; adopting risk management best practices for strategic readiness; and leveraging data science and artificial intelligence (AI) to better understand strategic readiness levels and forecast future ones.

DoD's Initial Implementation of the Strategic Readiness Paradigm

In 2023, DoD adopted the strategic readiness framework in an official DoDI to "inform senior leaders of the readiness trade-offs and impacts resulting from their decisions to better illuminate associated risks to force, mission, and the NDS [National Defense Strategy]."[2] For the first time, DoD defined *strategic readiness* as "the ability to build, maintain, and balance warfighting capabilities and competitive advantages that ensure the DoD can achieve strategic objectives across threats and time horizons."[3] This definition explicitly codified ten dimensions that capture the facets of strategic readiness in the department: operational

---

[1] Department of Defense Instruction (DoDI) 3000.18, *Strategic Readiness*, U.S. Department of Defense, November 30, 2023.

[2] DoDI 3000.18, 2023, p. 9.

[3] DoDI 3000.18, 2023, p. 17.

readiness, sustainment, mobilization, modernization, global defense posture, force structure, resilience, human capital, allies and partners, and business systems and organizational effectiveness.

This new DoD definition of strategic readiness was significant for department deliberations because it expanded the scope of military readiness beyond the immediate time horizon in two ways. First, it explicitly identified readiness as a quality to be tracked and planned for conflicts in the future time horizon. Readiness could also potentially be sought by accepting fewer objectives (ends) or less readiness (ways and means) in the immediate time horizon. Second, DoD readiness dimensions now measure a *transition into war* context and the ability to fight a protracted war. Operational readiness thus became a subset of strategic readiness, necessary but not sufficient for department readiness for conflict. Strategic readiness as a concept was meant to bring together the activities of many stakeholders—stretching from the defense industrial base to the training base to frontline forces—that support warfighting across the short-, medium-, and long-term time horizons.[4] The culmination of the DoDI for strategic readiness was the creation of a Strategic Readiness Assessment (SRA) with designated offices of primary responsibility for the ten dimensions to annually assess and report their trends and risks.[5]

Since publication of the Strategic Readiness DoDI, the Office of the Assistant Secretary of Defense for Readiness (OASD[R]) has increasingly been asked to conduct risk assessments of strategic readiness impacts from out-of-cycle increases in security assistance and global force employment. Senior leader demand for tailored, issue-specific strategic readiness assessments increased beyond expectations. Additionally, the staff effort and time required for data collection and validation of issue-specific assessments, such as security assistance munitions, has limited the number of senior leader questions that OASD(R) could address. Several risk decisions did not seem to match articulated DoD priorities, particularly in force employment of aircraft carrier strike groups or air defense batteries. Furthermore, developing DoD-wide integrated mitigation options for these specific issues and risk decisions proved to be an additional challenge.

Confusion persists throughout DoD about how the ten dimensions fit together, what should be measured, and what constitutes a risk that should be raised to senior leaders. Thanks to representation in several DoD processes (global force management, the Defense Management Action Group, the Defense Workforce Council, and budget review), OASD(R) leadership noticed that leaders in these processes were repeatedly confronting the same strategic readiness issue (e.g., defense industry production) from different angles but with little coordination of an overall mitigation approach across DoD processes and agencies. Depart-

---

[4] Stephen Watts, Ashley L. Rhoades, Michael E. Linick, Katharina Ley Best, Josyln Fleming, Paul W. Mayberry, John C. Jackson, and Michelle D. Ziegler, *Rethinking Strategic Readiness: A Framework for a Strategic Readiness Assessment*, RAND Corporation, RR-A2060-1, May 22, 2024.

[5] Deputy Secretary of Defense (DSD), "Implementing a Strategic Readiness Approach," memorandum, May 13, 2022, Not available to the general public.

ment leadership also observed that data points not typically considered in readiness reporting (e.g., defense industry production obstacles, safety and mishap incidents, experience levels of the aircraft maintainer workforce) appeared to reflect deeper strategic readiness trends, but analytically assessing those connections was difficult. For the annual SRA, it was not clear what the department should do with its findings to begin addressing risks—should it influence front-end guidance, spur deeper portfolio reviews on trending gaps, or generate mitigation tasks? Strategic readiness concerns were real, but DoD did not appear to have a consistent method or organizational capacity for turning potential leading indicators into data-informed assessments of downstream impacts.

Research Questions and First Principles

We were asked to develop ways to ensure that department decisionmaking regarding strategic readiness is pedagogically coherent, objective, informed by data, and based on risk. By *pedagogically coherent*, we mean an approach that is logically organized, aligned, and easy for practitioners to understand. Pedagogical coherence helps practitioners make connections between components, which allows them to act on those connections with inputs and activities to achieve desired outcomes.[6] We began our research with three lines of inquiry:

- Is DoD's conceptual frame for strategic readiness sufficient for operationalization in DoD processes?
- How does DoD manage strategic readiness risks across its key processes?
- How does DoD see its strategic readiness risk profile using the data and reports it already generates?

We grounded our research in two *first principles* on military readiness and defense planning. First, Richard Betts's classic framing remains as relevant today as when he wrote it almost four decades ago: Military readiness is essentially about answering the questions of "ready for what (what threat)," "ready with what (what forces)," and "ready when (what time frame)."[7] Strategic readiness for DoD must answer these same questions, but it must extend the scope of "ready with what" beyond warfighting capabilities for a particular adversary to the foundational systems (e.g., human capital systems, defense industrial base, training base) that sustain those warfighting capabilities for a protracted conflict or for decades of strategic competition. Second, Colin Gray captures the essence of the policymaker dilemma when he repeatedly emphasizes that defense planning is inherently risk management of different

---

[6] Adapted from Tim Oates, *A Cambridge Approach to Improving Education: Using International Insights to Manage Complexity*, Cambridge University Press, 2017, pp. 12–14.

[7] Richard K. Betts, "Military Readiness: Concepts, Choices, Consequences," Brookings Institution Press, February 1, 1995.

trade-offs and choices—about threats to prepare for, forces to build, and timelines for building those capabilities—in an inescapable competitive environment wrapped in uncertainties and contingencies that can upend expectations.[8]

Method

The starting point for our research was the 2024 RAND report *Rethinking Strategic Readiness: A Framework for a Strategic Readiness Assessment.*[9] The authors of that report critiqued pre-2023 DoD readiness assessments for having a narrow scope and short-term focus, using simplified measures, and taking a piecemeal approach.[10] To overcome these limitations, those authors put forward an understanding of strategic readiness as an "ecosystem in which a wide variety of current readiness metrics reside but that encompasses both a wider array of requirements and the interrelationships between them."[11] In particular, they flagged the need for *strategic framing* ("identifying the desired defense capabilities"), *sufficiency analysis* (working backward from desired defense capabilities to determine the joint force required), and *process monitoring* (monitoring DoD's "strategic readiness-generating processes to determine whether they can deliver the capabilities required on the necessary timelines").[12]

We embrace the *Rethinking Strategic Readiness* approach to strategic readiness assessments and expand on its thinking by (1) operationalizing the concept of strategic readiness and providing an alternative logic model for how the strategic readiness dimensions relate to each other, (2) discussing how to improve DoD processes so that these assessments have maximum impact when decisions are being made, and (3) showing how AI tools can help conduct the assessments faster and with less manual labor so that they can be generated in a more timely manner that allows for maximum impact at various decision points.

We began by reviewing the different definitions of readiness that DoD uses. For instance, there are individual levels of readiness, along with higher-echelons-of-command readiness. Next, we were interested in how the department monitors and manages risk within the ten dimensions of strategic readiness. Finally, we were interested in how DoD sees itself in terms of strategic readiness, especially in terms of the data, models, and systems that the department uses to assess strategic readiness. For each of these lines of inquiry, we took a similar methodological approach. We began with a review of literature internal to DoD but also in academic journals, gray literature, and other industry reports. We then reviewed definitions that the department uses for relevant readiness concepts and their monitoring processes.

---

[8] Colin Gray, *Strategy and Defence Planning: Meeting the Challenge of Uncertainty,* Oxford University Press, 2016.

[9] Watts et al., 2024.

[10] Watts et al., 2024.

[11] Watts et al., 2024, p. 5.

[12] Watts et al., 2024, pp. ix, 94.

We conducted semi-structured interviews with DoD subject-matter experts to document challenges and map processes related to strategic readiness. We also conducted a red team panel with RAND subject-matter experts on DoD readiness. Last, we identified gaps between best practices identified in literature and DoD practices to explore potential opportunities to build on the concept of strategic readiness.

Literature Review

We conducted an extensive literature review to establish how the concept of strategic readiness has developed thus far (discussed in Chapter 2), identify best practices in risk management (discussed in Chapter 3), and understand the strengths and weaknesses of various analytic tools (discussed in Chapter 4). Our literature review included prior RAND research; peer-reviewed, academic sources; books from academic thought leaders and practitioners on the history and nature of risk; risk management publications by consulting firms and business magazines; non-DOD government sources; and DoD sources.

Our textual and thematic analysis of DoDIs, Department of Defense Directives (DoDDs), Chairman of the Joint Chiefs of Staff (CJCS) Manuals and Instructions, U.S. Code, and other classified government instructions informed our understanding of the process gaps and bureaucratic hurdles facing implementation of strategic readiness activities and our understanding of the unevenness of emphasis among the ten strategic readiness dimensions. We also reviewed DoD and Joint Staff readiness assessments for the past two years, including the SRA, Cumulative Impacts to Strategic Readiness, the Chairman's Risk Assessment (CRA), and quarterly readiness reports to Congress.

Interviews with Stakeholders and Red Team with SMEs

The concept of strategic readiness is a relatively new doctrinal term in DoD, and many concepts and definitions associated with it are in nascent stages. To understand the department's thinking and way ahead for strategic readiness as a guiding principle in DoD systems and processes, we chose to engage in semi-structured interviews with Office of the Secretary of Defense (OSD) leadership and staff, as well as Joint Staff. We developed a semi-structured interview protocol after conducting a literature review of DoD internal documents and an academic and gray literature review. Our interview protocol focused on challenges and opportunities OASD(R) faces while integrating strategic readiness into contemporary systems and processes, such as planning, programming, budgeting, and execution (PPBE); global force management; and strategy and force development processes.

We held semi-structured interviews with dozens of subject-matter experts and senior leaders across OASD(R), including the Assistant Secretary of Defense of Readiness; the Deputy Assistant Secretaries of Defense for Force Readiness, Force Education and Training, and Force Safety and Occupational Health; and action officers in those offices. We also interviewed Joint Staff action officers who develop the CRA and Global Force Management Action Plan (GFMAP). Throughout the course of the project, we also engaged with the sponsor team in OASD(R) to provide interim project reports. In these meetings, we briefed the sponsor on key findings at each stage in the research process. During these meetings with leadership and staff, we held open discussions that informed our way ahead for subsequent interviews and analysis.

We also held a red team meeting with RAND subject-matter experts to solicit input on the progress of our research, identify areas of weakness within our research, and highlight areas needing additional work. Members of the red team possessed DoD and service expertise in strategic and operational military readiness, strategic and operational decisionmaking, military force planning, risk characterization and management, and logistics and resource management. The red team meeting was conducted on Microsoft Teams and used the software application Mural as a platform to allow for anonymous, group collaboration and simplify the data collection from multiple subject-matter experts.

Experimentation with Large Language Models and Unstructured Readiness Data

RAND has extensive experience analyzing data and developing dashboards, models, and simulations for DoD. Since 2020, data analytics has increasingly happened in cloud-based environments, such as DoD's Advana and Palantir Foundry systems branded as Army Vantage and Air Force Envision. RAND researchers and data scientists have also experimented with collection tools and techniques broadly known as AI, which in the past few years has been nearly synonymous with the blossoming capabilities of large language models (LLMs). RAND's work has focused on one or more dimensions of strategic readiness and has been conducted at both the unclassified and classified levels. In addition to being practitioners, RAND subject-matter experts have also examined such questions as how DoD can use AI and other analytic methods and why AI projects sometimes fail. We drew on all this experience, as well as the previously mentioned literature review and interviews with DoD stakeholders, to inform our analysis of how DoD can better see itself with data, analytics, and AI tools.

The Urgency of This Study

The U.S. defense establishment is now widely acknowledged as facing the most dangerous, uncertain, and technologically fast-evolving security environment it has seen in the life span of the republic—or, at the very least, since the end of the Cold War. Risk management of strategic readiness distills the essence of the policymaker dilemma in these challenging times. This dilemma is made more acute by the complex bureaucratic systems that are necessary for administering modern institutions but also make shifting course, gaining alignment, or coordinating efforts difficult. This report will show that clear conceptual frameworks, sound bureaucratic practices, and the ability to rapidly synthesize and dialogue with data indicators are critical means for policymakers to navigate through this uncertain security environment.

CHAPTER 2

Conceptualizing Strategic Readiness

Getting the massive DoD bureaucracy to adopt new business practices is always a challenge. With implementation of the strategic readiness framework, this challenge is amplified by the lack of shared understanding of what strategic readiness and its ten dimensions mean. This problem is exacerbated by the fact that many strategic readiness terms (readiness, mobilization, sustainment, modernization, resilience, human capital, allies and partners) predate the inception of strategic readiness and are defined differently by different parts of the defense enterprise. DoD definitions of readiness, in particular, have evolved over the past several decades. Our interviews found that, though there was general understanding of the definitions for strategic readiness dimensions (with a few exceptions), stakeholders did not have a shared understanding for how those definitions fit together from inputs to outputs and then to outcomes.

When dealing with complex systems, logic models are useful for linking objectives, outputs, activities, and inputs to well-designed measures that can inform decisionmaking about policies and resources.[1] This chapter will analyze challenges in operationalizing the definition of strategic readiness. We will then provide a logic model that brings together national-level policy guidance and DoD resource choices. The purpose of the recommended logic model is to provide an operationalizable framework for risk management of strategic readiness.

Readiness: Multiple Definitions, but One Definition to Bind Them All?

DoD organizations employ several definitions for *readiness*, which have evolved over time in answer to specific organizational imperatives (see Table 2.1). The embrace of strategic readiness in addition to military readiness is a relatively recent phenomenon within DoD, reflecting recognition that maximizing the pursuit of short-term operational readiness could have negative trade-offs for long-term readiness in the future because of deferred investments. The concept of strategic readiness begins to respond to the classic critique raised by Richard K. Betts that pursuing military readiness can detract from other priorities, such as developing

---

[1] Scott Savitz, Miriam Matthews, and Sarah Weilant, *Assessing Impact to Inform Decisions: A Toolkit on Measures for Policymakers*, RAND Corporation, TL-263-OSD, July 25, 2017.

and fielding new platforms, investing in research and development, and providing time for personnel and equipment to rest and refit.[2]

Different Readiness Definitions Across DoD

It is natural that different organizations or echelons within the U.S. national security apparatus define readiness differently based on their scope of responsibilities. However, the broad range of definitions demonstrates the difficulty with defining readiness and the risks of using the term *readiness* to inform DoD decisionmaking when readiness is viewed so differently by key actors (see Table 2.1).

Readiness definitions frequently come with their own tailored readiness metrics. DoD readiness metrics—such as the Defense Readiness Reporting System (DRRS), the Chairman's Readiness System, Quarterly Readiness Report to Congress, mission-capable rates, and aircraft-availability rates—provide only insights to specific facets of readiness. For example, DRRS only looks at readiness from a unit level, which does not automatically aggregate to the readiness of department mobilization and sustainment systems for conflict. Assessments using these metrics are incomplete for understanding the broader perspective on DoD readiness for conflict because they tend to focus specifically on combat unit readiness, short-term preparations, and a handful of scenarios.[3]

Strategic Readiness: A More Comprehensive Definition, But How to Operationalize?

In 2023, DoD for the first time explicitly defined strategic readiness as "the ability to build, maintain, and balance warfighting capabilities and competitive advantages that ensure the Department of Defense can achieve strategic objectives across threats and time horizons."[4] The purpose for this framework was to "inform senior leaders of the readiness trade-offs and impacts resulting from their decisions to better illuminate associated risks to force, mission, and the NDS."[5] DoD further identified ten dimensions that capture the facets of strategic readiness in the department: operational readiness, sustainment, mobilization, modernization, global defense posture, force structure, resilience, human capital, allies and partners, and business systems and organizational effectiveness. *Operational readiness*—often referred to as simply *readiness*—is thus a subset of strategic readiness focused on select military forces for today's assigned missions. Though essential, operational readiness is not sufficient to represent the entire department's readiness for conflict. Strategic readiness as a concept brings together the activities of many stakeholders—stretching from the defense industrial base to

---

[2] Betts, 1995.

[3] Watts et al., 2024.

[4] DoDI 3000.18, 2023, p. 17.

[5] DoDI 3000.18, 2023, p. 9.

TABLE 2.1

DoD Definitions of Readiness

Document	Proponent	Date	Term	Definition
CJCS Guide to the Chairman's Readiness System	CJCS	November 14, 2010	Readiness	"The ability of U.S. military forces to fight and meet the demands of the [National Military Strategy]"
CJCS Guide to the Chairman's Readiness System	CJCS	November 14, 2010	Readiness from the strategic perspective	"The ability of the joint force to perform missions and provide capabilities to achieve strategic objectives as identified in strategic level documents (National Security Strategy, National Defense Strategy, NMS)"
Army Regulation (AR) 525-30	Army	April 9, 2020	Army strategic readiness	"Army's ability to provide adequate forces to meet the demands of the NMS. It is measured quarterly through the Army Strategic Readiness Assessment (ASRA) process (see fig 1–2) utilizing one Army and three Joint Staff mandated assessments to obtain an integrated view of current and future strategic readiness."
DoDD 7730.65	OUSD(P&R)	May 31, 2023	Strategic readiness	"The requisite military power arrayed across time and space to achieve strategic ends. Looks beyond operational readiness of military forces to include supporting and complementary elements necessary to meet the defense objectives identified by the NDS."
10 U.S.C. § 117	Congress	Fiscal year 2024	Readiness	"The capability of the armed forces to carry out (1) the National Security Strategy . . . (2) the defense planning guidance provided by the Secretary of Defense . . . and (3) the National Military Strategy prescribed by the Chairman of the Joint Chiefs of Staff"
DoDI 3000.18	OUSD(P&R)	November 30, 2023	Strategic readiness	"The ability to build, maintain, and balance warfighting capabilities and competitive advantages that ensure DoD can achieve strategic objectives across threats and time horizons"
DoD Dictionary of Military and Associated Terms	CJCS	April 2024	Readiness	"The ability of military forces to fight and meet the demands of assigned missions"

SOURCES: AR 525-30, *Army Strategic and Operational Readiness*, U.S. Army, April 9, 2020, p. 2; CJCS, "CJCS Guide to the Chairman's Readiness System," November 15, 2024, p. 1; DoD, *DoD Dictionary of Military and Associated Terms*, March 2017, p. 195; DoDD 7730.65, *DoD Readiness Reporting System*, U.S. Department of Defense, May 31, 2023, p. 14; DoDI 3000.18, 2023, p. 17; U.S. Code, Title 10, Section 117, Readiness Reporting System.

NOTE: OUSD(P&R) = Office of the Under Secretary of Defense for Personnel and Readiness.

the training base to frontline forces—that support warfighting across the short-, medium-, and long-term time horizons.[6] The culmination of the DoDI for strategic readiness was the creation of an SRA with designated Offices of Primary Responsibility for the ten dimensions to annually assess and report their trends and risks.[7]

DoD's conceptualization of strategic readiness has two main issues. First, it refers to many complex topics without clear definitions. What is meant by *warfighting capabilities* and *competitive advantages*? What are *strategic objectives across threats and time horizons*? Second, the relationships between these complex topics are also unclear. How does building and maintaining warfighting capabilities and competitive advantages relate to strategic objectives across threats and time horizons? How do inputs into strategic readiness dimensions lead to outputs for warfighting capabilities and competitive advantages? This initial definition of strategic readiness, though succinct, needs more unpacking to be useful.

Of course, there is no such thing as a perfectly representative concept, but, above all, it should provide coherent cause-effect logics and facilitate operationalization through metrics and resource levers.[8] Interviewees observed that practices for characterizing strategic readiness are more art than science. The lack of clarity about how the ten strategic readiness dimensions relate to each other did not lend itself well toward analyses focused on achieving a specific goal or objective. Interviewees noted that the ten strategic readiness dimensions as currently framed do not facilitate systems optimization analysis or trade-off comparisons.

A well-operationalized concept also supports consistent, repeatable, timely analysis across different decisionmaking processes. Interviewees noted that the ten strategic readiness dimensions are not consistently presented in various options analysis. One individual noted that strategic readiness dimensions tend to recede into the background toward the tail end of decisionmaking in strategy development, PPBE, and global force management.[9] Another individual observed that analyses of strategic readiness impacts are often late-to-need for key decisions. Although OASD(R) produces an annual Strategic Readiness Assessment, interviewees noted that it would be helpful to have analyses more closely tied to in-stride decisions that affect strategic readiness, such as unplanned contingency operations deployments.[10] When emergent requests for forces arise, OSD and the Joint Staff often need weeks to gather data, decide how to frame the issue, and then synthesize insights from the services on risk to forces and risk to mission.[11]

---

[6] Watts et al., 2024.

[7] DSD, 2022.

[8] A concept has greater internal coherence the more its phenomena are logically and functionally related to each other. A concept is better operationalized the easier it is for observers to determine its borders and measure its phenomena (John Gerring, *Social Science Methodology: A Unified Framework*, 2nd ed., Cambridge University Press, 2011).

[9] Subject-matter expert, interview with the authors, March 21, 2024.

[10] Subject-matter expert, interview with the authors, March 18, 2024.

[11] Subject-matter expert, interview with the authors, March 2024.

Getting conceptualization right matters because it shapes how many different actors in the institution will approach and solve a problem. Senior leaders intuitively recognize that strategic readiness is a process of balancing risks and resource trade-offs across time horizons, not a specific end state to be achieved.[12] Balancing requires constant adjustments and trade-offs; it is never-ending and requires consistent application across key DoD processes. But how can senior leaders weigh these trade-off options with rigor and discuss them in different processes if leaders do not share a common conceptualization? These shortcomings lead to our first recommendation: Operationalize strategic readiness in a logic model that illustrates how inputs and activities lead to outputs and outcomes meant to meet strategic objectives across time.

Recommendation: A Logic Model for Strategic Readiness

Although the strategic readiness definition has proven an excellent starting point for providing clarity about the pieces that are important to strategic readiness, it lacks clear definitions of how those pieces relate to each other and produce desired outcomes. A logic model can help decisionmakers understand how a decision that they thought affected only one piece of the puzzle affects other pieces of the puzzle. It shows where unintended effects are most likely and helps get them out of a siloed perspective of the decision at hand. In short,

> a logic model typically offers a simplified visual representation of the path of a program's operations, starting with inputs and then progressing to the program's activities, its outputs, its customers, and its intended outcomes. The model may also link the program's operations, what the program actually does either alone or with others to fulfill its mission, to its strategy.[13]

A logic model should be a conceptual binder that illustrates the key relationships between inputs, activities, outputs, and desired outcomes or objectives.[14]

---

[12] Kimberly Jackson and David H. Berger, "Readiness Redefined: Now What?" *War on the Rocks*, June 12, 2023.

[13] Victoria A. Greenfield, Valerie L. Williams, and Elisa Eiseman, *Using Logic Models for Strategic Planning and Evaluation: Application to the National Center for Injury Prevention and Control*, RAND Corporation, TR-370-NCIPC, March 30, 2006, p. 3.

[14] *Rethinking Strategic Readiness* (Watts et al., 2024) offered a prototype logic model for strategic readiness. It showed how the ten strategic readiness dimensions relate to each other and how these functioned as inputs that were turned into outputs. In this prototype framework, sustainment, modernization, mobilization, and human capital all feed into each other and operational readiness, global force posture, and force structure. Operational readiness, global force posture, and force structure feed into military outcomes. Resilience, allies and partners, and business processes enable and mitigate military outcomes and sustainment, modernization, mobilization, and human capital. The outputs were military outcomes, which led defense outcomes and future readiness implications. The weakness of this prototype model is that it does not distinguish between warfighting capabilities and competitive advantages, it does not present a clear

Building Strategic Readiness: The Strategic Readiness Pyramid

We started building our logic model by unpacking DoD's strategic readiness definition and ten dimensions.[15] The first challenge was describing how these dimensions interact. In our logic model, we categorize the ten dimensions into warfighting capabilities and competitive advantages. *Warfighting capabilities* are things (forces, materiel, resources, basing, logistics, infrastructure) that constitute tangible military power for use in military operations. Warfighting capabilities have varying stages of preparedness: from execution-ready to on-hand. *Competitive advantages* are systems that generate more or better warfighting capabilities. By themselves, competitive advantages cannot be employed in military operations to achieve strategic objectives. Over the course of great power competition, competitive advantages are the key to transforming inputs (raw material and people) into warfighting capabilities. Competitive advantages can be broken down into *middle-term* and *long-term*.

Execution-ready warfighting capabilities (the green triangles in Figure 2.1) are ready and positioned for immediate employment today (sometimes called *fight tonight*) and include operational readiness and global force posture. Operational readiness should be scoped to the force's ability to execute a specific conflict scenario at that moment—how prepared it is to execute likely military tasks for that scenario with the personnel, equipment readiness, supply, and training it has at that point in time.[16] Global force posture should be scoped to whether sufficient forces, equipment, supplies, and basing infrastructure are positioned in the right location at that point in time for the forces to prevail. Together, operational readiness and global force posture determine what warfighting capabilities are immediately available for use in a specific scenario at a specific location. All other dimensions are inputs for this tip of the spear of execution-ready warfighting capabilities that executes the offensive, defensive, and stabilization missions that achieve strategic objectives.

On-hand warfighting capabilities (the blue triangles in Figure 2.1) are existing forces, equipment, and material that could be brought to bear against a threat with additional resources and time. On-hand warfighting capabilities are captured by force structure and sustainment. *Force structure* is the Total Force: active-duty, Reserve, and National Guard force elements. *Sustainment* encompasses the global quantities of war reserve materiel, replacement parts, supply stocks, lift assets, and distribution nodes. On-hand warfighting capabilities are all existing things that—with the investment of additional time, training, equipping, refit, or strategic lift—can be deployed to a theater as *execution-ready warfighting capabilities*. On-

---

trace of inputs to outputs, and it does not provide a frame for the trade-off discussion of the opportunity cost of actions today (Watts et al., 2024).

[15] DoDI 3000.18, 2023.

[16] Likely military tasks for a specific conflict scenario can be well defined through Joint Mission Essential Task Lists submitted by combatant commanders or projected based on Defense Planning Scenario analysis. In both the near-term and future cases, there is an element of judicious planning for "be prepared to" tasks that might arise.

FIGURE 2.1

DoD Strategic Readiness Profile: A Pyramid Logic Model

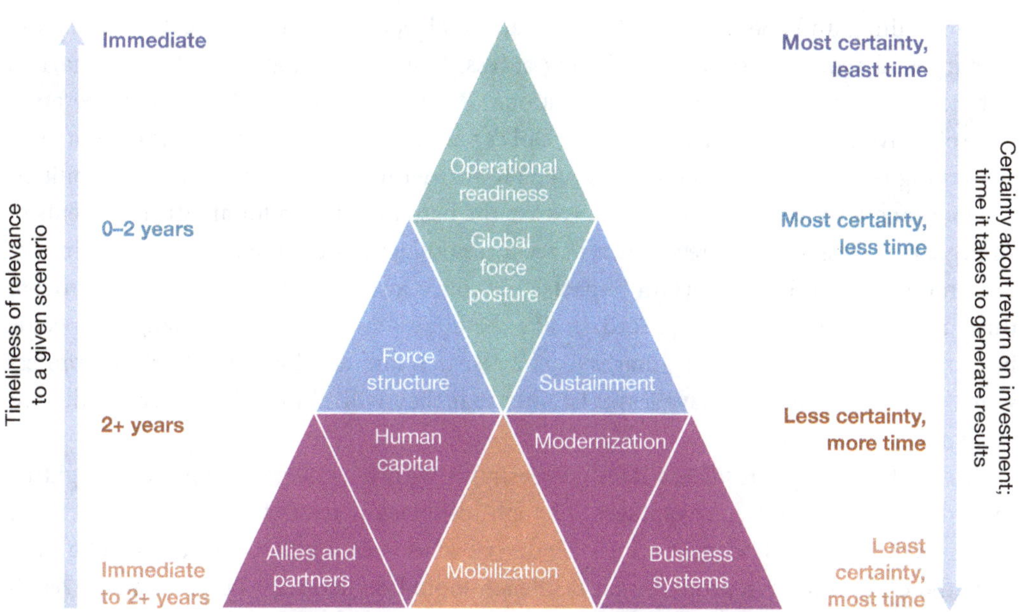

NOTE: Though resilience is treated as one of ten strategic readiness dimensions in DoD Instruction 3000.18 (2023), it is more accurately described as a feature that undergirds the other nine. Resilience is embedded in each strategic readiness dimension's ability to withstand, fight through, and recover quickly from disruption.

hand warfighting capabilities typically begin flowing into a theater of war within weeks of conflict initiation but can take up to a couple years to generate.

The *mid-term competitive advantage* focuses on mobilization (the orange triangle in Figure 2.1)—the ability and capacity to convert on-hand, domestic sources of power into warfighting capabilities for a protracted conflict. Mobilization depends on two pillars: the defense industrial base and the service training base. The defense industrial base transforms material stocks (e.g., semiconductors, critical minerals, steel) and fiscal resources into equipment and munitions. The service training base transforms recruits, activated reserve elements, and reconstituting active-duty units into trained, deployable forces. *Mobilization*, in this definition, is thus more than the typical limited conception of the process of bringing Reserve Component forces into active duty and preparing them for their mission. Force structure, sustainment, and mobilization typically determine what warfighting capabilities could be available to joint force commanders during the first two to three years of a conflict.

Long-term competitive advantages (the purple triangles in Figure 2.1) are the strategic readiness dimensions that generate joint force advantages over adversary warfighting capabilities over the course of multiple decades of competition. Long-term competitive advantages include the strategic readiness dimensions of human capital, modernization, allies and partners, and business systems. *Human capital* is about DoD recruitment, development, and retention of talent. *Modernization* concerns the investment in research, development, testing, and evaluation of better technologies, as well as investment in procuring and fielding capability sets to the force in a manner that enables the U.S. military to gain and maintain warfighting advantages. *Allies and partners* concerns developing relationships with partner nations that are willing and able to support U.S. strategic objectives in the political, information, military, and economic spheres, including access, basing, overflight, intelligence-sharing, technical cooperation, and force contributions. *Business systems* is about DoD operating more effectively and efficiently, increasing DoD's pace of building warfighting capabilities or conserving resources that can then be applied to further increasing warfighting capabilities. Notably, long-term competitive advantages are the base foundation for all other elements of strategic readiness but require significant resource outlays to develop; rarely provide immediate reward; result in less-certain impacts; and, once lost or degraded, require significant time and resources to redevelop. And, unlike the upper- and middle-tier triangles, specific investments in competitive advantages (e.g., a research and development effort or a professional education program) have less reliability that they will ultimately deliver the desired warfighting capabilities.

DoD's definition of *resilience* does not easily fit within a categorization of warfighting capabilities or competitive advantages. Though resilience is treated as one of ten strategic readiness dimensions, it is more accurately described as a feature that permeates and undergirds the other nine. We propose that resilience should be defined as each strategic readiness dimension's ability to "withstand, fight through, and recover quickly from disruption."[17] This definition aligns with the 2022 NDS idea of *deterrence by resilience*: discouraging adversary actions by reducing vulnerabilities across critical DoD systems. For example, resilience in sustainment could mean more-diverse supply chains or finding substitutes for necessary parts. Resilience in global force posture could mean hardening base infrastructure against attack or preparing emergency plans for bases susceptible to natural disasters.

These nine-plus-one dimensions form a strategic readiness pyramid that culminates in execution-ready warfighting capabilities at the point of need and constitute the department's strategic readiness profile (see Figure 2.1). Operational readiness and global force posture are immediately relevant to warfighting. Force structure and sustainment are relevant for reinforcing, resupplying, or reconstituting warfighting capabilities within the first months to two years of a given scenario. Mobilization is generally relevant after the first six months (more often at least a year) to regenerate forces or build new ones. Human capital, modernization, allies and partners, and business systems are relevant throughout the life cycle of a great power competition but generally take years to translate into significant warfighting capabilities. Resilience permeates the pyramid, helping DoD to absorb and recover from shocks to any part of the pyramid.

The overall state of the strategic readiness pyramid—its gaps, shortfalls, and trends—constitutes the department's strategic readiness profile. A key part of deterrence cost-benefit calculations are the dimensions of countable things (warfighting capabilities)—operational

---

[17] DoD, *2022 National Defense Strategy of the United States of America*, 2022, p. 8.

readiness, global force posture, force structure, sustainment—and allies and partners to assess the local balance of forces, particularly for immediate deterrence in a specific context.[18] Competitive strategies typically focus on the undergirding systems—mobilization, human capital, modernization, and business systems—that generate asymmetric advantages within a set of warfighting capabilities.[19]

The strategic readiness pyramid also illustrates varying uncertainty about the speed and predictability of returns on investments. Investments made in improving operational readiness or global force posture have the most immediate, predictable effect on warfighting capabilities, but investments in global force posture and, especially, operational readiness are also the most transient and subject to decay. Investments made in improving force structure and sustainment often take longer to mature than investments made in operational readiness and global force posture. Investments in human capital, modernization, and allies and partners not only take longer to mature, but they also have a less certain payoff (e.g., uncertainty about whether a talent management reform will produce the desired outcome, whether a technology investment will mature, or whether a partner nation will help when needed). Over-reliance on near-term time horizon measures of readiness biases DoD's understanding of its overall readiness toward near-term strategic objectives. Yet, as the strategic readiness pyramid logic model illustrates, if the foundation of competitive advantages is not solid, warfighting capabilities in the future can deteriorate.

How Much Will Be Enough—Where and When? Red Versus Blue Yardsticks

Any discussion of strategic readiness must answer three key questions: Ready for what? Ready for when? And ready with what?[20] DoD's definition of strategic readiness implicitly acknowledges the centrality of answering "for what" and "for when" with its charge that DoD must ensure that it can achieve strategic objectives across threats and time horizons. Strategic readiness is built to serve strategic objectives in the near-term (next two years) through mid-term horizons (end of the five-year defense program) while positioning the joint force to achieve strategic objectives in the long-term future (two to three five-year defense programs).

Strategic objectives are the desired ends of policy or, perhaps more accurately, the desired future environment that policymakers aim to achieve. U.S. presidential administrations usually establish their strategic objectives in early pronouncements of intent, while their formal codification in the National Security Strategy (NSS) and NDS can take a year or more. Other sources of strategic objectives include Presidential Decision Memorandums or secretary of defense guidance. Congress also plays a role in establishing U.S. strategic objectives through

---

[18] Michael J. Mazarr, *Understanding Deterrence*, RAND Corporation, PE-295-RC, April 19, 2018, pp. 3–5.

[19] Thomas Mahnken, "Thinking About Competitive Strategies," in Thomas Mahnken, ed., *Competitive Strategies for the 21st Century: Theory, History, and Practice*, Stanford University Press, 2012.

[20] Betts, 1995.

legislation that dictates policy objectives for the executive branch, senate ratification of defense treaty commitments, and congressional allocation of resources.

The next step is to translate strategic objectives into red and blue yardsticks (Figure 2.2). The use of yardstick comparison tools for such dynamic portfolio management is well grounded in the literature of defense decisionmaking science and has been generally adopted in department plan reviews and future defense planning scenarios.[21] Each red yardstick represents the warfighting capabilities that an adversary will be able to bring to bear in a particular scenario at a particular time to achieve red strategic objectives. Red yardsticks represent

FIGURE 2.2

Execution-Ready Warfighting Capabilities to Strategic Readiness

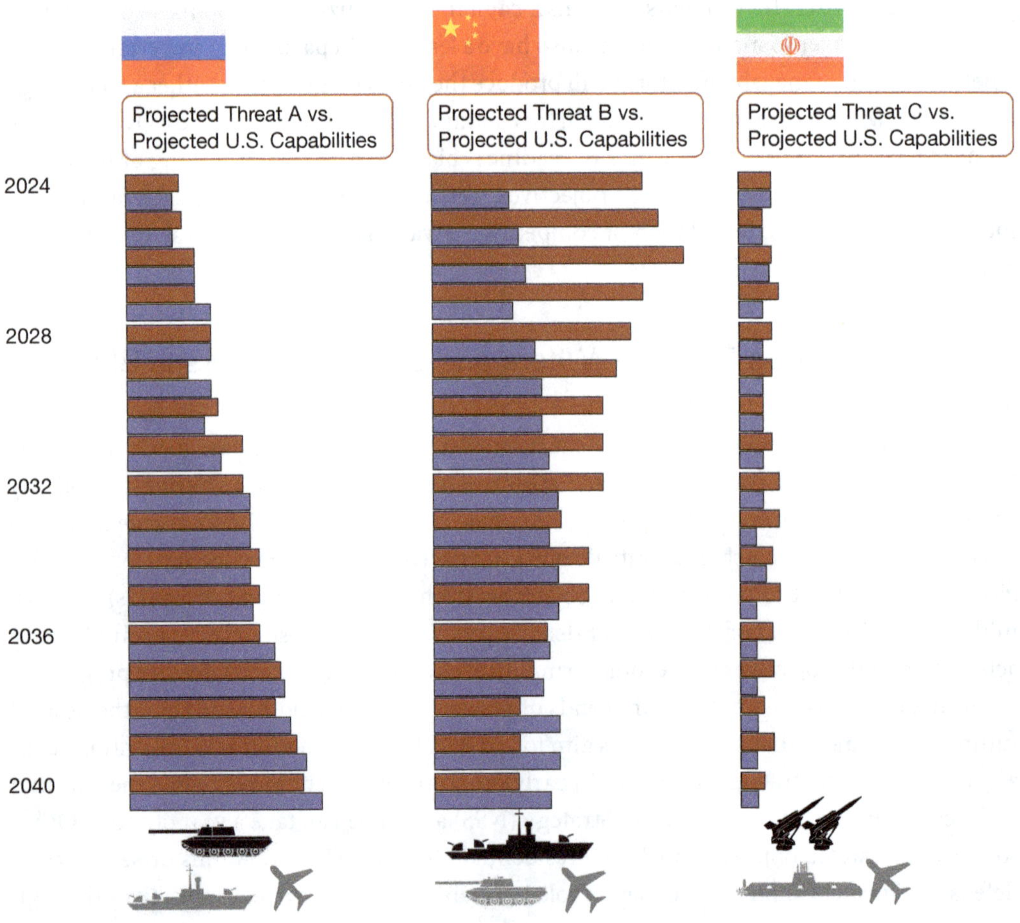

---

[21] Paul K. Davis, *Analysis to Inform Defense Planning Despite Austerity*, RAND Corporation, RR-482-OSD, March 20, 2014. See also Paul K. Davis, Jonathan Kulick, and Michael Enger, *Implications of Modern Decision Science for Military Decision-Support Systems*, RAND Corporation, MG-360-AF, 2005; and Paul K.

what blue must be able to overcome to achieve blue strategic objectives. Crude measures of red yardsticks could include projected red operational readiness, force structure, munitions, and equipment. Each blue yardstick represents the projected U.S. warfighting capabilities (ready with what) that could be brought to bear to achieve blue strategic objectives against a specific threat—threat A, B, or C (ready for what)—in a specific time horizon—2024, 2025, 2026, or beyond (ready for when). In this stylization, we chose not to reflect how different mixes of force types may be required for strategic objectives for A, B, or C, although this level of decomposition could easily be examined with land, air, sea, cyber, and space yardsticks.

Of course, projecting with absolute certainty what must be available to blue in 2026, 2027, 2028, or beyond against threats A, B, and C is not possible. Rather, the purpose of red yardsticks is to convey two key ideas. First, different strategic objectives for a threat have different warfighting capability requirements. Second, warfighting capability requirements needed to achieve strategic objectives against a given threat may increase over time (as with threat A), decrease over time (as with threat B), or remain relatively constant over time (as with threat C), depending on red force development, capability investments, or losses in other conflicts.

In this model, red yardsticks do not represent the probability that a conflict will erupt with red or the consequences of blue failure to achieve its strategic objectives. Nor does the model capture the impact that choosing not to meet threats B or C today would have on the likelihood and consequences of conflict with threats (A, B, and C) in the future. Probability and consequence trade-offs for near-term strategic objectives are better examined through the extensive body of intelligence assessment methodologies.[22] In the long term, projecting probabilities and consequences of threats to strategic objectives is sometimes better explored through forecasting methodologies.[23]

Red and blue yardsticks are also silent on different operational concepts that could be used, although warfighting capabilities are the most tangible and measurable building blocks of those operational concepts. Wargames and modeling are more appropriate analyses for exploring the comparative value of different operational concepts for a specific contingency

---

Davis, *Capabilities for Joint Analysis in the Department of Defense: Rethinking Support for Strategic Analysis,* RAND Corporation, RR-1469-OSD, 2016.

[22] See Steven W. Popper, "Reflections: DMDU and Public Policy for Uncertain Times," in Vincent A. W. J. Marchau, Warren E. Walker, Pieter J. T. M. Bloemen, and Steven W. Popper, eds., *Decision Making Under Deep Uncertainty: From Theory to Practice,* Springer, 2018. See also Paul Bracken, Ian Bremmer, and David Gordon, eds., *Managing Strategic Surprise: Lessons from Risk Management and Risk Assessment,* Cambridge University Press, 2008. See also Richard B. Heuer, *Psychology of Intelligence Analysis,* Center for the Study of Intelligence, 1999; Cynthia M. Grabo, *Anticipating Surprise: Analysis for Strategic Warning,* University Press of America, 2004; and Katherine Pherson and Randolph Pherson, *Critical Thinking for Strategic Intelligence,* Sage, 2013.

[23] See Peter Schwartz and Doug Randall, "Ahead of the Curve: Anticipating Strategic Surprise," in Francis Fukuyama, ed., *Blindside: How to Anticipate Forcing Events and Wild Cards in Global Politics,* Brookings Institution Press, 2007; and Maree Conway, *Foresight: An Introduction; A Thinking Futures Reference Guide,* Yuba Community College District, 2015.

plan or defense planning scenario. Indeed, wargames and modeling should inform estimates for blue yardsticks—recommending the quantity and types of forces needed to overcome red yardsticks. The proposed comparison of red and blue yardsticks is also not a substitute for in-depth net assessments that delve deeply into strategic cultures, asymmetric advantages, and long-term sources of power for the development of long-term, grand strategy. Instead, the purpose of the comparative red and blue yardstick approach is for strategy execution when an emergent policy choice could affect future commitments and resource availability.

Balancing Risk Is a Function of Two Conversations

Balancing strategic objectives and strategic readiness across time horizons is a continuous, ongoing process dependent on disciplined conversations at two levels: the national policy level and the institutional DoD level (see Figure 2.3). The national policy conversation revolves around interconnected choices about prioritizing threats and time frames, setting strategic objectives, developing ways (diplomatic, information, military, economic), and allocating resources. The institutional DoD conversation is about how to best apply DoD resources across time to support national policies (see Figure 2.4).

FIGURE 2.3

Strategic Readiness Profile: A Fulcrum Between Strategic Objectives and Military Means

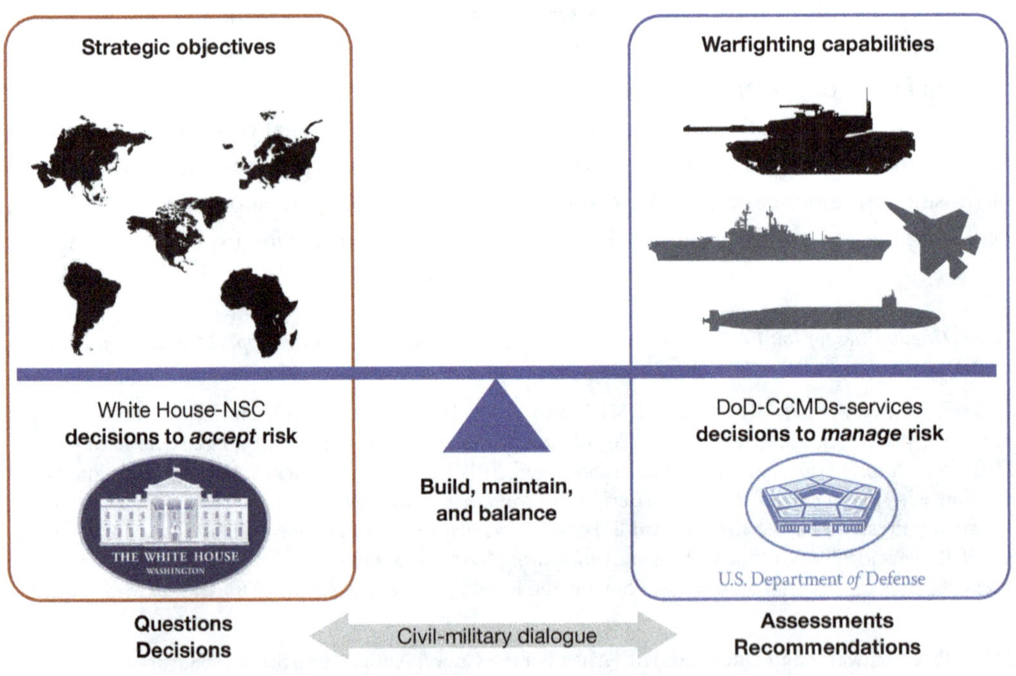

NOTE: CCMDs = combatant commands; NSC = National Security Council.

FIGURE 2.4

Execution-Ready Warfighting Capabilities to Strategic Readiness

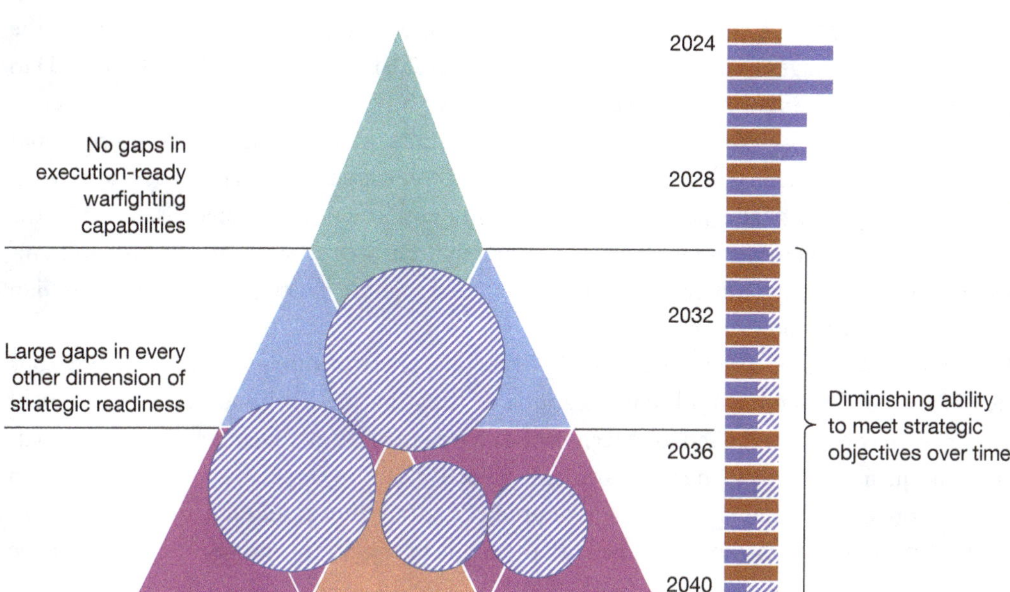

On the left side of the fulcrum in Figure 2.4 are the red yardsticks based on policy guidance—an accounting of what the U.S. military must be able to overcome to meet the strategic objectives set by policymakers for different threats across different time horizons. In the left-side external or national policy conversation, OSD and Joint Staff engage with the White House, Congress, and interagency members of the National Security Council to inform decisionmaking about the strategic readiness impacts of policy proposals. In particular, OSD and Joint Staff help explain trade-offs of policy proposals for other strategic objectives and threats, now and in the future. In the national policymaking process, OSD and Joint Staff should inform these deliberations with rigorous, data-informed, and risk-based analysis of downstream impacts to the strategic readiness pyramid. Ultimately, it lies with the White House to provide policy guidance and with Congress to provide resources.

On the right side of Figure 2.3 are the blue yardsticks of warfighting capabilities projected to be needed for assigned strategic objectives, built from DoD's allocation of investments, resources, forces, and policies across the strategic readiness dimensions. The right-side conversation is the internal DoD conversation—involving OSD, Joint Staff, services, CCMDs, and DoD agencies—about options for managing risks and closing gaps in the strategic readiness pyramid. DoD resource allocations shape the types and quantities of warfighting capabilities for strategic objectives for different threats (in this logic model, the length of the blue yardsticks). Today's commitment of forces or munitions stocks typically reduces downstream strategic readiness without the commitment of additional resources. If the pyramid has no gaps in execution-ready warfighting capabilities (the green triangles in Figure 2.4—

operational readiness and global defense posture) but large gaps in reinforcing force structure or sustainment stocks, then department warfighting capabilities will erode over time or diminish rapidly in a conflict (see Figure 2.4). Furthermore, if there are large gaps in the competitive advantages that form the base of the pyramid, then it will be difficult for DoD to keep pace in generating warfighting capabilities across future years.

Because DoD faces time and resource constraints, there will be trade-offs in choices about which gaps to close. This will mean accepting risk in some areas. Managing where and when to accept risk is a critical aspect of strategic readiness that will be further discussed in the next chapter. Big decisions about how the department pursues strategic readiness and manages risk are ultimately political decisions that, by design, have oversight from senior civilian defense appointees and Congress.

Of course, threat assessments of probability and severity of threats across time horizons are embedded throughout both conversations. Estimating the likelihood of a threat event occurring in the near term and long term is the provenance of intelligence agencies. Estimating consequences of failure to meet a strategic objective can be as much a political discussion of policy preferences as it is an intelligence prediction. In either case, evaluating DoD methods for future threat assessments is beyond the scope of this study and not necessary for the logic model.

Risk management of strategic readiness is essentially the process of trying to bring the left side—strategic objectives—and the right side—warfighting capabilities across time—in balance with each other (see Figure 2.3). The proposed logic model should make conversations about strategic readiness among senior political and military leadership more coherent by clarifying what the real stakes under consideration are in the following ways:

- For political leaders: How does making this policy decision (e.g., diplomatic commitment, force employment, force posture, budget changes) change the red and blue yardsticks for strategic objectives (means required)? How will this political decision affect the tilt in the U.S. military strategic readiness fulcrum (means available for the future)? How does not committing forces or resources—staying the course—affect the tilt in the strategic readiness fulcrum (threat metastasizing)?
- For military leaders: How does the decision at hand change the ability to meet blue yardsticks across time (means available)? How will this decision affect the strength and stability of DoD's strategic readiness pyramid (long-term health of institutional competitive advantages)?

The logic model can also help military leadership in the Joint Chiefs, CCMDs, and the services articulate to the Secretary of Defense and Congress where risk is accumulating in strategic readiness and where more resources could help.

Conclusion: Visualizing Strategic Readiness as a Risk Management Process

The purpose of this chapter was to provide a logic model of DoD's strategic readiness profile that can be used to inculcate a shared understanding of strategic readiness across DoD. This chapter used logic models that built on each other to illuminate how the strategic readiness dimensions connect in space and time to produce warfighting capabilities at a time of need, how strategic objectives influence strategic readiness requirements, and how policy decisions to commit (or not commit) U.S. military resources to strategic objectives affect strategic readiness in the future. The logic model proposed in Figures 2.1–2.4 shows visual heuristics, not an attempt to exactly capture reality. This logic model does not attempt to capture all the input-output interconnections between strategic readiness dimensions or display the level of detail necessary for implementing detailed action plans. The purpose of the logic model is to advance a common framework for discussing institutional risk management of strategic readiness.

There are three key takeaways regarding risk management that follow from the strategic readiness logic model. First, there is an inverse relationship between how foundational a strategic readiness dimension is to the entire DoD enterprise and how easy it is to quantify and measure it. In other words, the long-term competitive advantages are the most difficult to quantify, measure, and assess in terms of long-term impact on readiness. This raises a challenge: If DoD only focuses on strategic readiness dimensions that are easily quantified and measured, DoD may miss trends that are more foundational to the health of the entire enterprise; take longer to be recognized as problems; and, most glaringly, take the longest to rectify.

Second, there are two sides of the equation to balancing strategic readiness. The external conversation approaches this challenge by making conscious decisions about where to accept more risk in strategic objectives or by adjusting strategic objectives. These are political decisions that need to be deliberated by the White House, Congress, and the National Security Council in consultation with DoD leadership. The internal conversation approaches this challenge by closing gaps in warfighting capabilities and competitive advantages with a consideration for the stability and strength of the entire pyramid. These are decisions that need to be made by consultation among OSD, the Joint Staff, CCMDs, services, and agencies. Being strategically ready—finding the balance on the fulcrum between the warfighting capabilities and competitive advantages DoD has and the strategic objectives across threats and time horizons that DoD needs to be able to achieve—will require adjustments to be made through both the external and internal conversations.

Third and finally, because it is impossible to be perfectly ready for all threats across all time horizons, the best DoD can do is have the internal conversations to manage the risks that have been deemed acceptable by the external conversations. Strategic readiness is similar to risk management: Being ready for everything all the time is as impossible as eliminating all risks. Strategic readiness is a continuous process, not an end state. National policy leaders and DoD leadership will face difficult decisions. Lower-priority strategic objectives might be deprioritized to improve the odds against higher-priority strategic objectives. Some gaps in warfighting capabilities and competitive advantages might be accepted to close even larger or more foundational gaps. Although these are difficult decisions, it is better that the decision be made consciously within a framework that reveals trade-offs and mitigates risk. By making acceptance of the gap a discrete risk decision, it allows for focusing some effort on ways to address the risk accepted. What can be injected from the decision into processes for budget, force employment, plan development, or concepts and force modernization to address the gap in the future?

Having a shared, sound organizational conception for how warfighting capabilities are generated over time is only a first step for grappling with strategic readiness. Risk management also depends on consistent organizational practices and sound knowledge, assessments, and forecasts. Chapter 3 will recommend ways to practically operationalize the strategic readiness logic model within DoD by evaluating DoD processes and comparing them with risk management best practices. Chapter 3 will also explore ways to better conceptualize department strategic readiness through emerging AI tools that improve how DoD analyzes strategic readiness impacts and monitors strategic readiness trend lines.

Risk Management of Strategic Readiness

In this chapter, we discuss how DoD manages risk to its strategic readiness profile in three parts. First, we review best practices for risk management in large institutions operating in the risk-laden sectors of finance, insurance, and resource extraction. Analyzing different concepts of risk and how industries approach and manage it suggests ways DoD could improve its risk analysis, risk deliberations, and risk mitigations. Second, we examine DoD practices for risk management of the department strategic readiness profile. Third, we provide recommendations for how to improve DoD processes by incorporating industry best practices.

Best Practices for Risk Management from Finance and Industry

The section below will look at how industry defines risk, views risk exposure, and mitigates risk. We begin by acknowledging that *risk* is a loaded concept with layers of meaning. The word *risk* originated in 16th-century Italy in tandem with the systematic assessment of probability theory.[1] Risk theory leapt forward at the beginning of the 20th century as mathematicians, philosophers, and economists began to explore the concept of risk and its implications for decisionmaking. Risk has come to mean multiple things to different organizations with implications for how to manage or mitigate it, but one aspect of risk that remains consistent over time is its inextricable link with probability.[2]

Risk Management Best Practices in Finance, Insurance, and Resource Extraction

The finance, insurance, and resource extraction industries must manage significant risk in a high-stakes environment where getting risk management wrong can lead to institutional failure. Like DoD, these industries are often global in scope, can be overwhelmed by liabilities, and come under intense government scrutiny. Extractive industries that invest in a particu-

---

[1] Peter L. Bernstein, *Against the Gods: The Remarkable Story of Risk*, Wiley, 1998.

[2] Michael J. Mazarr, *Rethinking Risk in National Security: Lessons of the Financial Crisis for Risk Management*, RAND Corporation, CB-549, June 1, 2016.

lar resource can be locked into that investment for years, or even decades, and they also face workforce casualties and physical security threats. These industries start from similar definitions of risk as a function of the probability of an event combined with its impact on a thing of value. These industry definitions of risk mirror the CJCS's definition of risk in the Joint Risk Analysis Methodology (JRAM) ("the probability and consequence of an event causing harm to a thing that is valued").[3] Considering these industries allows us to survey a landscape of risk factors with similarities to DoD and draw on their risk management approaches for potentially adoptable practices.

Financial investors seek to maximize gain while keeping risk at or below their risk profile (the level of risk they are willing and able to accept). Financial investors face two main types of risk: market and idiosyncratic. *Market risk* underscores that there is no such thing as a risk-free investment; participating in the competitive market is inherently risky. *Idiosyncratic risk* is risk to the specific asset or investment (e.g., investing in soybean futures exposes the investor to risks from drought, global trade disruptions, changes in diet preferences, and more). Investors cannot get rid of market risk, which is why it is important to have a risk profile. Investors try to minimize idiosyncratic risk through hedging and diversification. Insurance companies face risk from insuring high-risk populations or insuring a single market that exposes them to either corrosive incremental losses or a large-scale disaster. Investors use actuarial science to minimize initial risk exposure, then purchase reinsurance. Though it does not remove the risk, reinsurance distributes the risk more broadly throughout the financial system, thereby making it less likely that, for example, a catastrophic flood would lead to financial collapse for the reinsurance company or leave the initial insurance companies unable to pay out claims. Resource extraction industries contend with an array of risks as they make large capital investments in countries threatened by war, revolt, or authoritarian caprices. They use a variety of mitigation strategies, including investing in a variety of geographic regions, investing in different types of resources, diversifying their customer base, using technology to minimize environmental harm, conducting geopolitical risk analysis, and engaging in rigorous safety training protocols.[4]

The common thread of successful industry risk management is a disciplined risk management process and internal control system integrated across the organization. Industries seek to manage risk by understanding their exposure to risk (through developing a risk profile, conducting actuarial science, or conducting geopolitical risk analysis) and distributing risk throughout a broader ecosystem (through diversification of investments, hedging, and reinsurance) so that they are not overexposed to a single risk vector. These institutions often also maintain procedures, monitoring systems, and regular touchpoints with leadership to

---

[3] CJCS Manual 3105.01B, *Joint Risk Analysis Methodology*, December 22, 2023, p. B-1.

[4] Mazarr, 2016. See also Martin Pergler, "Enterprise Risk Management: What's Different in the Corporate World and Why," McKinsey, 2012; Kevin Buehler, Andrew Freeman, and Ron Hulme, "The Risk Revolution," McKinsey, 2008a; and Kevin Buehler, Andrew Freeman, and Ron Hulme, "The New Arsenal of Risk Management," *Harvard Business Review*, September 2008b.

prevent risk from accumulating within month-by-month operations. Finally, they develop operational approaches, training protocols, and safety standards to not only minimize operational risk but also be postured to deal with potential negative outcomes when they happen. At the heart of risk management approaches are senior leadership conversations about the willingness to accept risk, the ability to accept risk, and the mitigations that leaders choose to pursue.[5]

Reviewing industry risk management case studies and RAND research on risk reveals that implementing a disciplined, integrated risk management process has several necessary institutional conditions:

- *A shared appreciation for risk:* The first quality of effective risk management is an organizational culture that encourages candor by allowing open dialogue, supporting dissent, and seeking alternative "what if" perspectives. A shared culture is strengthened by a shared appreciation for how risk accumulates and aggregates within an institution. This shared understanding is more than just common definitions of risks as functions of probability and consequences or identification of risk factors. A truly shared appreciation is usually based on a well-understood institutional model for how inputs, outputs, and risks in one facet of the business affect other business operations, especially the end product.[6]

- *Independent oversight function:* An effective best practice is the use of an independent overseer to coordinate and overwatch the organization's internal risk control system and facilitate risk conversations. One of the most important roles of independent oversight is providing early warning of emerging or accumulating risk—particularly when a proposed policy or development in the environment increases the likelihood or severity of potential negative outcomes in the future. To be effective, this institutional risk oversight body must participate in all key business processes (e.g., strategy, financial, operations, distribution, sales).[7]

- *Investing in data ecosystems and tools that build transparency about risks within the organization:* Transparency means sharing the data between different business units that reveal risks running through the organization. Transparency in data and key metrics resists bureaucratic tendencies to silo information or oversimplify it as it advances up the chain of responsibility. Transparency also allows other business units to become aware of emerging risks sooner and be proactive to adjust their actions or take mitigation steps. A key condition for transparency is a common understanding of how inputs, activities, and outputs of all the core business units build the desired outcome. Objec-

---

[5] Mazarr, 2016.

[6] Robert S. Kaplan and Anette Mikes, "Managing Risks: A New Framework," *Harvard Business Review,* June 2012b; Robert S. Kaplan and Anette Mikes, "JP Morgan's Loss: Bigger Than 'Risk Management,'" *Harvard Business Review,* May 23, 2012a. See also Buehler, Freeman, and Hulme, 2008b.

[7] Kaplan and Mikes, 2012b.

tive, auditable risk-monitoring tools build confidence in their assessments and allow better identification of new metrics to be tracked as the environment evolves and new decision dilemmas emerge.[8]

- *Engaged senior leadership:* Senior leaders set the tone when they encourage alternative viewpoints and prioritize investigating serious, data-based warnings.[9] But culture alone is insufficient. Institutions that successfully weather risk usually have established "a regular cadence of conversations around risk, driven by those at the top of the organization."[10] Senior leaders must drive forums and mechanisms for bringing risks forward into their strategic decisionmaking, including regular touchpoints to review the institution's comprehensive risk profile as the environment changes. Senior leadership must also provide clarity about their prioritization of objectives and their risk thresholds. Thresholds are a revealing framework that empowers subordinates to flag accumulating risks for senior leader attention.[11]

- *Consistent follow-through:* Risk management is incomplete without consistent follow-through on mitigation actions across core institutional processes. Even if an informed decision is made to accept a certain risk, the institution should coordinate actions to mitigate that risk and monitor for it. The full life cycle of risk management includes a risk mitigation and monitoring plan that establishes thresholds for when the risks that were once accepted should be raised for senior leadership consideration again.[12]

Existing DoD Practices for Risk Management of Strategic Readiness

As we saw in Chapter 3, although multiple definitions of readiness exist within the department, all readiness definitions eventually are measured against the White House's NSS via the Secretary of Defense's NDS and the CJCS's National Military Strategy. These strategic documents provide front-end planning guidance for the five major department processes that represent the primary levers for risk management of strategic readiness dimensions. These are (1) Cost Assessment and Program Evaluation (CAPE) and service-managed budget processes (PPBE); (2) OSD-directed development of contingency, campaign, and posture plans by CCMDs; (3) the Joint Staff–led force employment process (global force management);

---

8 Mazarr, 2016.

9 Buehler, Freeman, and Hulme, 2008b.

10 Kevan Flanigan and Adam Regelbrugge, "Enterprise Risk Management (ERM): The Modern Approach to Managing Risks," Deloitte, September 25, 2023.

11 Flanigan and Regelbrugge, 2023; Volkan Evrin, "Risk Assessment and Analysis Methods: Qualitative and Quantitative," Information Systems Audit and Control Association, April 28, 2021.

12 Martin Pergler and Eric Lamarre, "Upgrading Your Risk Assessment for Uncertain Times," McKinsey, January 2009; Evrin, 2021.

(4) the service- and OSD-led modernization processes of research and development, requirements, and acquisition; and (5) the OSD-led Defense Management Action Group forums for making policy changes. These five processes are supported by two baskets of assessment processes for knowing the adversary and knowing ourselves. The Intelligence Community produces periodic, recurring threat assessments and specific-issue threat analyses to inform the five core processes. The services and Joint Staff monitor a myriad of metrics and produce a myriad of assessments about the readiness and state of the joint force (see Figure 3.1).

At the DoD level, although there is an overarching risk assessment framework for strategic readiness, there is not an overarching risk management process for strategic readiness.[13] There is a huge distinction between integrating reporting and integrating mitigation responses; statutory directives and DoD organizational structures are designed for the former, not the latter. U.S. Code charges the Secretary of Defense with establishing a comprehensive readiness reporting system to "measure in an objective, accurate, and timely manner the capability of the armed forces to carry out" the President's NSS, the NDS, and the National Military Strategy.[14]

In DoD processes, the strategic objectives of the NSS and NDS are translated into specific military objectives that guide readiness through two classified policy directives: Guidance for the Employment of the Force (GEF) and Defense Planning Guidance (DPG). The GEF focuses on force employment over the next two to four years, providing specific military campaign objectives for each region. The GEF typically includes Contingency Planning Guidance (CPG) with specific military objectives for designated conflict contingencies. The GEF and CPG are wholly policy directives, provided by the Secretary of Defense with the approval of the President. DoD refreshes the GEF and CPG, at a minimum, with every four-year administration and new NDS, although it has occasionally been updated around the middle point of an administration. The DPG focuses on force development for the services, providing specific guidance for their development of warfighting capabilities across a Fiscal Year Defense Program. The DPG typically designates select defense planning scenarios as yardsticks for service warfighting capabilities for the 5- to 15-year time horizon. The Secretary of Defense issues the DPG annually for each budget cycle. The GEF, CPG, and DPG are "north star" internal DoD policy directives that provide measurable strategic objectives to drive DoD force employment and force development of warfighting capabilities.

This Office of the Under Secretary of Defense for Policy (OSD[Policy])-led strategy process to articulate balanced strategic objectives into the middle- and long-term horizons is heavily deliberated and reviewed by service and CCMD stakeholders. However, the next step

---

[13] See DoDD 5105.79, *DoD Senior Governance Framework*, U.S. Department of Defense, November 8, 2021. See also DoDI 3000.18, 2023. DoD-level risk management frameworks typically focus on specific issues, such as acquisition program risk or cybersecurity network risk (DoDI 8510.01, *Risk Management Framework for DoD Systems*, U.S. Department of Defense, July 19, 2022). See also DoD, *Risk, Issue, and Opportunity: Management Guide for Defense Acquisition Programs*, September 2023.

[14] 10 U.S.C. § 117.

FIGURE 3.1
Five Core DoD Processes

NOTE: DAS = Defense Acquisition System; DRT = Directed Readiness Tables; GFMAP = Global Force Management Action Plan; JCIDS = Joint Capabilities Integration and Development System; PPBE = planning, programming, budgeting, and execution; RDT&E = research, development, test, and evaluation; SDOB = Secretary of Defense

of identifying strategic readiness requirements, or even operational readiness requirements, in those time frames is underdeveloped and has no OSD proponent. Historical experience has shown that DoD's defense planning scenario process is an imperfect vehicle for determining actual force requirements; it is designed to test the robustness of proposed force structures against several illustrative scenarios and is purposefully kept separate from near-term operational plan (OPLAN) requirements.[15] OSD CAPE often finds itself in the position of articulating de facto force requirements for purely budget review purposes.

OSD oversight and policy responsibilities for readiness fall to the Under Secretary of Defense for Personnel and Readiness (USD[P&R]). USD(P&R) is designated the principal staff assistant to the Secretary of Defense for readiness, Total Force management, and training. USD(P&R) responsibilities include developing policies, plans, and programs for readiness to "ensure forces can execute the National Military Strategy."[16] USD(P&R) is also responsible for oversight of DRRS. U.S. Code Title 10, Section 117 directs that DRRS should measure unit readiness, training establishment capability, defense infrastructure capability, and critical warfighting deficiencies. DRRS focuses on real-time reporting of near-term unit readiness—personnel levels, training, equipment serviceability, and material supplies.

CJCS has an essential advisory role in reporting and assessment of readiness risks. Over the decades since Goldwater-Nichols reform, the Joint Staff has developed and refined an in-depth system for tracking readiness of force elements. The CJCS's JRAM defines risk as "the probability and consequence of an event causing harm to a thing that is valued."[17] The JRAM conceptualizes a risk management process of four steps: (1) identify risk, (2) assess risk, (3) mitigate risk, and (4) make decisions that balance risk cost with mission benefits. The JRAM manual's portrayal of this process includes the importance of problem framing, gathering knowledge of risks, application of judgment about the risk, and communication of those risks to higher echelons of leadership. However, the JRAM is focused predominantly on the first two steps: identifying risks and assessing risks.

CJCS develops risk assessments of readiness through the CRS, which integrates service DRRS-based assessments, as well as OPLAN assessments and joint capability area assessments—again, with a pronounced focus on near-term assigned missions. Besides regularly advising the Secretary of Defense on risks to joint force readiness, CJCS submits two periodic risk assessments to Congress: a quarterly force readiness assessment (known as the Joint Forces Readiness Review) and the annual CRA, which assesses risks against objectives and ways of the National Military Strategy. One of the few explicit mentions of risk mitiga-

---

[15] Michael J. Mazarr, Katharina Ley Best, Burgess Laird, Eric V. Larson, Michael E. Linick, and Dan Madden, *The U.S. Department of Defense's Planning Process: Components and Challenges*, RAND Corporation, RR-2173/2-A, 2019, pp. 33–35.

[16] U.S. Code, Title 10, Section 136, Under Secretary of Defense for Personnel and Readiness. See also DoDD 5124.02, *Under Secretary of Defense for Personnel and Readiness*, U.S. Department of Defense, June 23, 2008, p. 3.

[17] CJCS Manual 3105.01B, 2023, p. B-1.

tion planning in Title 10 is indirectly connected to the CRA, when Section 153 directs the Secretary of Defense to include a plan for mitigating risks identified as significant or higher by CJCS.[18]

The Senior Readiness Oversight Council (SROC) and Executive Readiness Management Group (ERMG) are the principal DoD governance mechanisms for readiness. The SROC is chaired by the DSD with senior officials from the services and Joint Staff. The SROC largely focuses on review of quarterly and semi-annual readiness reports from the services and Joint Staff but also occasionally considers special readiness topics.[19] ERMG responsibilities are scoped to recommending policies and procedures for oversight and management of a comprehensive readiness reporting system, with special emphasis on DRRS.[20] The ERMG is chaired by the Assistant Secretary of Defense for Readiness and the Director of the Joint Staff.[21] The narrowly scoped responsibilities of these supporting governance forums suggest that their focus is on quarterly and semi-annual readiness reports—readiness of select force unit elements and joint force readiness for OPLANs—and oversight of the systems for collecting, assessing, and reporting that readiness data.

Over the last two decades, DoD undertook a major transition from separate Joint Staff and DoD service databases to one common, real-time DRRS database. Beyond integrating DoD operational readiness reporting systems, DoD has worked to bring together more than 1,200 internal data systems into the Advana platform, with potential profound implications for tracking strategic readiness.[22] Advana took a fragmented collection of unstandardized data and interfaces and meshed them together in a tool that allowed the data to be viewed in simplified dashboards for easier analysis. As its usefulness was quickly proven for financial auditing, additional data and functions were added to use the tool in other ways.[23] In May 2021, DSD Kathleen Hicks noted that Advana was the "single enterprise authoritative data management and analytics platform" for DoD. Advana is envisioned as bringing disparate data together in one place to exist as the authoritative "single source of truth" that would assist leaders in making objective, informed decisions.[24] Instead of having to consult various

---

[18] U.S. Code, Title 10, Section 153, Exchange of Material and Disposal of Obsolete, Surplus, or Unclaimed Property. The JRAM classifies CRA risks as military strategic risks (threats to U.S. interests), military risks (threats to mission execution and support), risks to mission (ability to execute a mission), or risks to force (ability to generate forces, sustain force health, and develop the force).

[19] DoDD 5149.2, *Subject: Senior Readiness Oversight Council (SROC)*, Office of the Director of Administration and Management, July 23, 2002.

[20] DoDD 7730.65, 2023.

[21] DoDD 5105.79, 2021.

[22] Grace Lin, "Meet Advance: How the Department of Defense Solved Its Data Interoperability Challenges," *Government Technology Insider*, April 7, 2021.

[23] Sydney J. Freedberg, Jr., "CDAO Opens Advana to Analytics to Multiple Vendors in a Push to Scale Up," *Breaking Defense*, July 9, 2024.

[24] Kathleen Hicks, "Creating Data Advantage," memorandum, U.S. Department of Defense, May 5, 2021.

offices in charge of many different business systems to get the data they need to make decisions, decisionmakers could consult Advana to get the data they need quicker and in such a manner that additional analytics can be run. DoD described its vision for Advana as a tool to "change decisionmaking behavior across the DoD enterprise using data and analytics."[25] As DoD's approach to managing data, analytics, and AI evolves, so too does its organization. Along with the release of the first AI strategy, 2018 saw the creation of the Joint Artificial Intelligence Center and the Office of Advancing Analytics, which managed Advana.[26] The use of Advana and trust in its data remain a work in progress as it scales up to serve more purposes.[27]

However, although the department has updated its reporting systems and data collection for readiness several times since 2001, the department has not significantly updated its governance structures for readiness since 2002.[28] Notably lacking from DoD charters for readiness governance is a department-level role in assessing strategic readiness trade-offs for key decisions, developing risk mitigation planning, and monitoring implementation of chosen risk mitigation actions. U.S. Government Accountability Office (GAO) reports in 2019 and 2023 noted that the department has no mechanism for DoD to examine readiness beyond a service-centric lens, particularly for routinely assessing joint force needs, cross-domain implications, and force structure trade-offs.[29]

In 2023, the OASD(R) began to address these gaps in risk assessment of strategic readiness. The department created the annual SRA to look beyond operational readiness and provide readiness assessments of sustainment systems, mobilization systems, modernization, human capital, department business systems, and allies and partners.[30] The SRA was an initial step to expand the aperture of department assessment of Total Force readiness to inform department-wide strategy and resource prioritization. Soon department leadership turned to OASD(R) to assess the strategic readiness implications for emerging policy choices for security assistance to Ukraine and additional force deployments to the Middle East. Working with the services and Joint Staff, OASD(R) developed several bespoke assessments. The Security Assistance Readiness Impact Analysis examines the impact of drawing down equipment and munitions from service inventories on the ability to execute future OPLANs or meet

---

[25] Greg Little, Jerimiah Bennett, and Kevin Winnike, "Advana Procurement Data," U.S. Department of Defense, 2022.

[26] DoD 7000.14-R, *Department of Defense Financial Management Regulation*, November 2020.

[27] Anastasia Obis, "DoD Preparing to Recompete for Advana," *Federal News Network*, July 16, 2024; Lauren C. Williams, "Pentagon Pauses Development of Its Go-To Data Analytics Tool," *Defense One*, June 10, 2024.

[28] DoDD 5149.2, 2002.

[29] GAO, *Military Readiness: Improvement in Some Areas, but Sustainment and Other Challenges Persist*, GAO-23-106673, May 2, 2023.

[30] DSD, 2022.

training requirements. The Cumulative Impacts to Strategic Readiness assessment periodically reviews impacts from GFMAP decisions.

Concurrently, OASD(R) launched an effort to develop decision-support tools to illuminate potential impacts on future force availability. OASD(R) is building the Readiness Decision Impact Model (RDIM) to provide data-driven forecasts of the planned supply of ready forces. OASD(R) also developed the Force Risk Reduction tool to gather injury, mishap, and safety data from across the department to help the services and OSD proactively identify trends eroding equipment and operational readiness. The intent of these tools is to support department risk management of strategic readiness to be more anticipatory of future impacts and proactive in development of mitigation actions. However, although the department has devoted significant energy to expanding its assessments of readiness risks and developing new tools for forecasting readiness risk, it has made almost no changes to its governance and processes for managing those readiness risks.

Evaluating DoD Against Best Practices from Risk Management for Industry

If DoD's prime execution function is winning the nation's wars across time, then DoD's prime management function is managing the dimensions of strategic readiness to ensure that the United States will be ready to prevail in conflict for its most important strategic objectives across time horizons. Viewing strategic readiness in time and space through the lens of risk management best practices illuminates several DoD shortcomings:

- *DoD conceptions and measurements of readiness focus on the near term.* DoD does not have a shared, DoD-wide framework for what *strategic readiness* means in practice (what DoD must be ready for when) and how to compare today's DRRS data with future readiness projections. The DRRS and the quarterly and semi-annual readiness assessments are a sound foundation for tracking near-term readiness risks, but they have shortcomings for broader application to DoD. The DRRS risk perspective predominantly audits operational readiness and force structure risk; it is not designed for monitoring risks in sustainment, modernization, or human capital systems. The nascent SRA attempts to fill this gap but lacks sufficient data and rigor because it relies on offices of primary responsibility to gather issues from stakeholders.
- *DoD lacks a dedicated advocate for conserving today's warfighting capabilities for the mid-term future.* The services attempt to fill this role, but they face countervailing pressures to maximize their operational employment of forces in the near term as justification for force structure capacity or to not be the service that says "no" to a force request. DoD-level approaches to risk management of strategic readiness tend to be focused on generating front-end guidance or delivering end-of-period reporting assessments.
- *Governance siloes and competing priorities inhibit a holistic risk picture.* Strategic readiness dimensions are managed through siloed governance processes for plans, budget, force employment, and acquisition—which are buffeted by competing stakeholder

interests and time horizon priorities. Moreover, OSD, Joint Staff, services, and CCMDs oversee different processes that generate decisions about risk but whose implications are not well communicated or integrated into other processes. The siloed governance structure makes it difficult for different components of DoD to understand how the issues they are focused on are affected by decisions made by other DoD components. Key governance decisions are not captured, documented, and reviewed through governance processes in a manner that provides for consistent understanding of risks and impacts to strategic readiness across time horizons.

- *There are missing conversations about the risk trade-offs of national policy decisions.* Department governance of readiness risk does not have clear channels for addressing the other half of the strategic conversation about risk—the external conversation about how to prioritize, balance, and mitigate risk among strategic objectives in time. Part of the reason for this gap is that these are policy decisions about how to prioritize competing national interests—decisions ultimately residing with civilian political leadership. Core mechanisms for the external conversation are the national policy deliberation process and legislative resource allocation process. Although CJCS and Joint Staff are key participants and advisors, it is civilian political leadership that drives national policy deliberations in terms of setting their issue agendas, raising second- and third-order implications for national interests, and framing trade-offs.[31]

- *Accumulated risks from incremental decisions are not well understood.* Many incremental decisions are made in process stovepipes without a chance to review the full, accumulating impact on hidden force readiness (e.g., safety, training, maintenance and sustainment stocks, budgetary impacts on modernization). In many areas where risk accumulates over time, there is no advocate or mechanism to raise the issue to the DoD level in a timely manner.

- *Risk mitigation is not integrated across services and agencies.* Risk is managed through several governance forums that are only truly integrated at the DSD or Secretary of Defense level. In practice, combined OSD and service actions are essential to addressing second- and third-order hazards to generating warfighting capabilities over time. Yet, DoD lacks a process or mechanism for developing integrated department mitigation actions and controls. When strategic readiness risk and impact mitigation occurs, it is often considered through the narrow lens of a particular platform, OPLAN requirement, or force element operational readiness rate.

- *OSD is not fully leveraging available data and reporting to identify strategic readiness risk.* Data from personnel training records, unit training evaluations, joint training exercise reports, maintenance and part failure rates, and injury and mishap statistics could have value as leading indicators of emerging strategic readiness risks. On the other hand, OSD struggles with the quality of data from less-federated data sources—particularly

---

[31] The Joint Strategic Planning system is designed for generating advice, plans, and assessments to communicate to higher levels.

those captured and transmitted via briefing slides. OSD also needs easier data access and more-sophisticated tools to support dynamic assessments of strategic readiness impacts across the middle term. Potentially rich sources of readiness insights are largely unexamined because they are captured in unstructured text (prose) in the myriad of reports and assessments produced by the Joint Staff and the services.

- *OSD lacks sufficient staff capability for integrating risk data, developing integrated depart-ment solutions, and following up with stakeholders on mitigations.* OSD and the Joint Staff should play important roles to integrate risks and facilitate external and internal risk conversations through agenda setting, stakeholder convening, and decision support. The Joint Staff has well-developed staff processes to support CJCS in these conversations, but OSD lacks a similar dedicated organizational function for integrating strategic readiness risk. OSD has the capability to conduct only cursory evaluations of massive annual planning products—such as the GFMAP and defense budget—for their impacts to the DoD strategic readiness risk profile. No one in OSD or the Joint Staff supervises or evaluates the efficacy of directed risk mitigation actions.[32]

Recommendations: Embedding Risk Management of Strategic Readiness in Department Governance

Risk management of strategic readiness is not an end state; it is a continuous process "to build, maintain, and balance warfighting capabilities and competitive advantages . . . [to] ensure [DoD] can achieve strategic objectives across threats and time horizons."[33] Risk management of strategic readiness is not a function that can be easily tacked onto existing portfolios. OSD, working with the Joint Staff, CCMDs, and services, is the appropriate staff body for teeing up issues for the national-level policy agenda, articulating risk trade-offs of policy options, and developing DoD-wide mitigation solutions. However, OSD lacks an organizational body and capabilities dedicated to tracking risks; raising warnings; developing integrated, resource-informed options; and tracking follow-through of mitigation measures. The

---

[32] Interviewees highlighted that the only mechanism that comes close is the OASD(R)-drafted Risk Mitigation Plan, which is formally charged with reporting to Congress on risk mitigation measures. But even this document is primarily designed to respond to risks identified in the annual Joint Staff–drafted CRA as opposed to being a real-time risk mitigation tracking tool. Statutorily, if the CRA characterizes overall baseline risk as "significant" or higher (and it always does), the Secretary of Defense is required to submit a Risk Mitigation Plan to Congress that identifies necessary adjustments to authorities, policies, priorities, operations, activities, and investments to address risks that are "significant" or higher. Theoretically, the Risk Mitigation Plan could constitute the missing step of "supervise and evaluate" risks, as the Risk Mitigation Plan is meant to feed back into annual DPG. But there is no formal mechanism mandating this, and the gap in supervising and evaluating risk remains.

[33] DoDI 3000.18, 2023, p. 17.

following recommendations focus on practical ways to create such an organizational body and capabilities.

Establish an Independent Integrator and Oversight Function for Strategic Readiness

The scale and scope of risk management for strategic readiness recommend the establishment of an independent oversight function within OSD that is dedicated to integrating across the five key decisionmaking processes: OSD-led defense policy review; service- and OSD-led modernization; OSD- and service-led PPBE; OSD-directed, CCMD-developed plan development; and Joint Staff–led force employment. This independent oversight function needs these authorities, relationships, and data access to drive integrated risk assessments, raise issues of accumulating risk, and coordinate risk mitigation across DoD.

Some might argue that CAPE already fulfills the role of independent integrator and oversight function for risk management strategic readiness. However, CAPE responsibilities (and organizational structure) focus predominantly on oversight of defense acquisition programs and annual program budget review. These two statutorily directed responsibilities require intensive leadership and analyst workloads inside CAPE.[34] Although CAPE regularly conducts strategic portfolio reviews on Secretary of Defense–directed topics as part of program budget review, these strategic portfolio reviews are carefully scoped and do not represent consistent long-term, holistic monitoring of all the facets of strategic readiness. In the past, CAPE has struggled to manage additional responsibilities beyond budget and cost assessments. During the 2010s, much of CAPE's analytic architecture for force planning scenario analysis was dismantled because of a lack of confidence in the work being produced.[35] CAPE's core stakeholder interest is resource management oversight. In this role, CAPE is an important line of defense for risk management of strategic readiness, but it can become compartmentalized from a broader perspective that interrogates long-term strategy risks or envisions emergent external risks.[36]

Multiple points in DoD governance processes could benefit from a broader independent integrator providing data-informed, integrated assessments projecting the impact of their decisions on future readiness. For example, when participating in front-end strategic guidance development—such as the development of the NDS, GEF, Global Force Management Implementation Guidance, or DPG—the independent oversight body could engage in dif-

---

[34] Michael Boito, Tim Conley, Joslyn Fleming, Alyssa Ramos, and Katherine Anania, *Expanding Operating and Support Cost Analysis for Major Programs During the DoD Acquisition Process: Legal Requirements, Current Practices, and Recommendations*, RAND Corporation, RR-2527-OSD, 2018.

[35] Eric V. Larson, *Force Planning Scenarios, 1945–2016: Their Origins and Use in Defense Strategic Planning*, RAND Corporation, RR-2173/1-A, 2019, p. 291. See also Mazarr et al., 2019.

[36] For the limitations of a resource-focused perspective, see Mazarr et al., 2019. For risk management best practices, see Kaplan and Mikes, 2012a.

ficult conversations with the drafters about trade-offs between the different directions that they are considering. Their purpose would not be to provide drafters with answers but rather to provide assessments that explicitly surface the implicit trade-offs among choices that the drafters are considering—trade-offs across threats, time horizons, and strategic readiness dimensions.

Similarly, DoD senior leadership who make decisions about force employment, security assistance, and foreign defense commitments would benefit from an independent staff body to integrate service, CCMD, and defense agency analyses for what the impacts of a potential decision will be on readiness in multiple time horizons, vis-à-vis other threats, and across other strategic readiness dimensions. These independent, integrated, data-based Total Force risk profile assessments need to be regularly integrated into decisionmaking for force employment and stockpile consumption, among other factors.

Finally, the independent integrator could provide a middle-term and long-term perspective when DoD officials are conducting GFMAP/SDOB risk reviews, program budget reviews, modernization and recapitalization planning, and warfighting concept reviews. The independent integrator can be charged to raise the alarm if a decision will have an unanticipated negative impact regarding DoD's readiness toward another threat not directly under consideration or in a strategic readiness dimension that does not fall under the purview of that particular decision's silo. An independent risk integrator function would be better positioned than current staff processes to

- maintain dedicated focus on Total Force strategic readiness for future wars beyond the immediate time horizon
- maintain a common operating of the holistic DoD risk profile and connect the dots across processes and stakeholders
- drive integrated assessments that provide insights on cumulative Total Force risks across the seams of time horizons, risk owners, and strategic objectives
- elevate strategic readiness risks in department processes through threshold monitoring, agenda setting, and independent assessments
- provide a neutral voice free of stakeholder resource interests that can raise risk warnings
- support senior leader decisionmaking with information on how different courses of action affect future strategic readiness across multiple dimensions
- serve as a focal point for staff support to senior DoD leadership for the external conversation with national political leadership about risk trade-offs
- track risk decisions and follow-through on risk mitigation actions.

These functions—the hallmark of risk management best practices—are best operationalized through a codified organizational responsibility with sufficient bureaucratic authority and a singular dedicated focus on monitoring the entire institution's risk profile. In DoD's case, the independent oversight function would participate in key department processes that make significant decisions—Defense Management Action Groups, plan reviews, program budget reviews, GFMAP and SDOB review—to help the DSD and Secretary of Defense. As an independent oversight function, decision support to senior leadership and stakeholders would be its primary mission, unencumbered by pressures to advocate for policies and resources for any particular DoD component.

In practical application, the independent oversight function would not have to require entirely new processes, reporting requirements, or governance bodies. For example, the existing SROC could be reinvigorated to raise emerging strategic readiness risk issues for senior leadership deliberation, built around working groups for the strategic readiness dimensions and leveraging existing assessment processes. The Defense Workforce Council governance stream could be used to monitor the Total Force risk profile in resiliency and human capital, integrating assessments of trends in personnel, health, safety, and training. Moreover, crossing a risk threshold could automatically elevate a mitigation choice within a set time frame through the existing Defense Management Action Group process.

Develop Clear, Executable Risk Guidance and Processes

Clear, executable risk guidance for strategic readiness requires clarity about senior leader risk concerns and suborganizational responsibilities. Developing risk guidance requires identifying (1) key decisions that would benefit most from a strategic readiness perspective and (2) leading indicators for when risk thresholds are being crossed. Key decisions that would benefit most from a strategic readiness perspective are those that have a complex and large impact on medium- and long-term readiness across multiple strategic readiness dimensions. This dialogue should be used to set priorities for developing data, tools, and assessments. Key issues that would benefit most from a strategic readiness perspective are those shaped by incremental decisions, those with long correction lead times, those that quickly transition from being a minor to major impediment to readiness, and those that do not already have a strong institutional champion.

OSD(Policy) and Joint Staff have a unique role to codify requirement yardsticks and risk thresholds for OPLAN failure risks. OSD and Joint Staff could further advance clear risk guidance by curating and codifying leading indicators for each of the ten strategic readiness dimensions. These OSD responsibilities should extend to oversight of service Title 10 measurements for manning, training, equipping, and Reserve component mobilization readiness. Examples of leading indicators include safety and operational mishap rates, unit training trends, personnel education and skills trends, force element dwell-to-deploy ratios, service equipment or munition production delays, and service maintenance backlogs.

A complementary effort to having clear risk guidance is clarity about risk responsibilities. Although the Joint Staff has a process map for risk assessment responsibilities in the Joint Strategic Planning System, OSD lacks a parallel process map for risk management responsibilities among the Under Secretary roles and DoD process owners. A codified DoD process map for how department processes connect with each other during deliberations—in particular, the points where integrated risk management review occurs—would create a more consistent, repeatable, and rigorous rhythm for an institutional review of Total Force strategic readiness trends. Developing clear executable risk guidance involves the following steps:

- OSD(Policy), in dialogue with DoD leaders and supported by Joint Staff, periodically revalidates defense strategic objectives across threats and time horizons.[37] The existing OSD strategy review group process could govern this task.
- OSD(Policy), CAPE, and Joint Staff, in dialogue with DoD leaders, estimate the red and blue yardsticks for the key strategic objectives. Source data for these yardsticks could include OPLANs, defense planning scenarios, and CAPE force assessment reviews. The existing Asymmetric Warfare Group process is well suited to govern this task.
- OSD(Policy), CAPE, Joint Staff, and services, in dialogue with department leaders, identify risk thresholds or zones of risk that should be prioritized for department tracking and review.
- A DoD process map is developed. Similar to the Joint Strategic Planning System, the process map would codify guidance inputs, decision outputs, cross-process integration points, organizational responsibilities, and risk management review points across the five key department processes for plans and strategy review, global force management, concept development and acquisition, program and budget development, and department policy forums.
- The existing SROC and ERMG governance processes are reinvigorated to serve as a periodic risk review board for the department's comprehensive strategic readiness risk profile—with particular focus on identified risk thresholds.

Develop Capabilities to Dynamically Track and Assess Strategic Readiness Risks

Even the best risk management practices are critically dependent on how well an organization understands itself. The challenge of visualizing an accurate, data-informed picture of military readiness has troubled the department since the 1960s.[38] In an institution as large as DoD with five services, 16 CCMDs, two dozen defense agencies, and two integrating staffs (Joint Staff and OSD), the sheer volume and mutability of data and assessments quickly become overwhelming. Typically, DoD has relied on a sequential hierarchy of assessments that move up the chain of command, which causes such challenges as latency of information, loss of nuance or key details, and potential for softening of conclusions or urgency.

Operationalizing a more rigorous understanding of DoD's strategic readiness risk profile will require further integrating department data, tools, and assessments. OSD and Joint Staff

---

[37] OSD(Policy) development of the NDS, GEF, and CPG essentially accomplishes this function, but this guidance may occasionally need update.

[38] Alain C. Enthoven and K. V. Smith, *How Much Is Enough? Shaping the Defense Program, 1961–1969*, RAND Corporation, CB-403, October 20, 2005.

have several ongoing initiatives to improve data sharing. The next step by OASD(R) has been developing prototype tools to integrate those data streams and make forecasts about strategic readiness impacts. However, department allocations of mid-level leadership, staff, and data science expertise are unable to keep up with assessment opportunities, especially for in-stride decisions.

Maturing advances in data analytics, modeling and simulation, and AI can help OSD better track strategic readiness. These tools can aid in analyzing the effects of decisions over time, promoting shared understanding of conditions, monitoring thresholds and critical metrics, and facilitating communication among senior leaders. AI is particularly useful for harvesting data across a myriad of sources and flagging potential problems. Most importantly, these AI tools are already giving general analysts the ability to *dialogue with the data*—by querying a wide range of sources, summarizing unstructured data, and identifying trends across data bases—in a way that has not been possible before. Three tools in particular have significant potential to enable an independent oversight function to better monitor, assess, and understand strategic readiness trends: dashboards, modeling and simulation, and AI assistants.

DoD already uses a proven set of DRRS dashboards to track operational readiness. Dashboards and other reports can be particularly useful for monitoring risk thresholds, as well as facilitating a shared understanding by making metrics visible to users and promoting transparency by drilling down into underlying data. However, DoD could do more to integrate data from all of the services, CCMDs, and defense agencies on areas outside operational readiness (e.g., joint training exercises, security assistance events, sustainment stocks, injury and mishap trends, recruiting and retention rates).

Computer modeling and simulation are widely used by DoD to address a variety of issues, from missile defense to modernization to personnel management.[39] Models are most useful because they provide a laboratory where variables can be manipulated, either alone or in groups, to see how different elements of the model affect one another and the model outputs. The greater the fidelity of those variables and elements to the real world, the better our understanding of how complex systems in the real world work. Models that examine slices of strategic readiness have proven more accurate and useful. The RDIM, developed and maintained by OASD(R), is an example of a simpler model that examines the relationship between units' deployment schedules and their operational readiness. Because of the modular nature of the model, it is relatively easy to add new factors, such as extreme climate events, that could also affect readiness.[40] It is now possible to start to link smaller models together via integration

---

[39] See, for example, Gary J. Briggs, *Harnessing Constructive Simulations for Reinforcement Learning*, RAND Corporation, RR-A1722-6, August 8, 2024; and Pete Schirmer, *Computer Simulation of General and Flag Officer Management: Model Description and Results*, RAND Corporation, TR-702-OSD, August 6, 2009.

[40] As this report was being written in 2024, another RAND research team was wrapping up a project to examine climate effects on training readiness and delivered new code to OASD(R) as an extension of the RDIM (Best, Katharina Ley, Scott R. Stephenson, Susan A. Resetar, Tim Wu, Peter Schirmer, Nathaniel

protocols, such as application programming interfaces, event cues, or standardized formats that can be used as inputs into federated models.

The growing capabilities of AI have the potential not only to supercharge both methods but also to offer new analytic tools to support risk management. A weakness of dashboards and models is that they are highly dependent on structured information that can be read from data tables, manipulated mathematically, put into charts, or distilled to metrics. Much of the information about strategic readiness is in the form of unstructured text. LLMs and their associated tools could harness the data in reports, assessments, and memoranda that are invisible to dashboards and models. LLMs can then be fine-tuned to perform tasks to better understand the specialized terminology of an organization, such as DoD, or categorize a domain, such as strategic readiness, or answer questions about a particular strategic readiness dimension. The development of retrieval-augmented generation tools in only the past two years allows an organization to store its proprietary or classified documents in a specialized database that allows a user to ask a question and generate answers from an LLM using those documents. In practical terms, it is possible to take a vast library of documents and ask questions: "How will expansion of force structure for the U.S. Air Force affect host nation support?" or "Which organizations raised readiness concerns about sharing Patriot air defense systems with Ukraine, and what did they say?" For example, commander comments in monthly DRRS reports could be a rich source of insights. LLMs can read those reports at scale, summarize and extract themes, and allow users to query them.[41] Furthermore, there are ways for AI to improve trust in the system, such as having an LLM automatically cite all sources and using a second LLM agent to cross check them. Some agents are designed to be much more transparent with their logic and the steps they take.

A concerted effort to build, focus, and empower analytic capabilities under an independent oversight function could drive integration of key data streams, accelerate development of tools, and provide a focal point for the community of interest that assesses and tracks readiness. Specific actions include the following:

- Gather and connect data streams in collaboration with the Chief Digital and Artificial Intelligence Office, the services, CCMDs, and DoD agencies.
- Issue stronger directives regarding data collection and data-sharing by services, CCMDs, and DoD agencies across a universally used department data system. Data directives should emphasize standards for ensuring the quality and consistency of the data, as well as expanding collection formats to gather data in the present that may be of value for future AI-enabled assessments.

Edenfield, Lewis Schneider, and Emmi Yonekura, *Climate and Readiness: Quantifying Climate Effects on U.S. Joint Force Training*, RAND Corporation, RR-A2921-2, forthcoming.

[41] In fact, RAND researchers built tools to read and interpret commanders' comments using an earlier generation of AI tools, as described in Peter Schirmer and Jasmin Léveillé, *AI Tools for Military Readiness*, RAND Corporation, RR-A449-1, September 20, 2021.

- Expand DoD ability to access and integrate non-DoD data for analysis.
- Build relevant tools to forecast strategic readiness impacts to inform key scheduled decisions in the department's five core processes, such as program budget review or the GFMAP. Transparency with services will be critical to foster open discussion, interrogate assumptions, and trace the methodology that led from the data to the projections.
- Work with services in the software development of tools (dashboards, algorithms, forecasting models, and visualizations) that help DoD leadership better see Total Force readiness across time and understand the impact of their decisions.
- Increase OSD staff capacity and grow OSD expertise to answer emerging senior leader questions using data streams from across the services, Joint Staff, and DoD agencies. As the RDIM matures, establish and promulgate standards with other offices within DoD to enable interaction between models that are being built, whether specifically about readiness or not.[42]
- Experiment with the use of LLMs to read and summarize readiness reports and other readiness-related assessments at scale—across stakeholders and looking back in time.
- Use the development operations approach of close collaboration by embedding tool designers with readiness analysts to modify and expand on existing tools.

Conclusion: Changing How DoD Makes Strategic Decisions

DoD intuitively weighs strategic readiness impacts when it makes decisions in its five core processes. However, the department does not apply a consistent, routine method when evaluating strategic readiness impacts or integrating mitigations across decision processes. Instead, risk management governance is siloed across different bodies focused on competing time horizons, accumulated risks and impacts from incremental decisions and actions are not visualized, multiple fora are largely integrated for the first time at the DSD level, and senior leaders lack dedicated staff capability for developing integrated department solutions and following up with stakeholders on mitigations. To address these shortcomings, we recommend that the department create an independent oversight function for the risk management of strategic readiness; develop clear, transparent risk yardsticks and thresholds mapped against DoD processes; and improve department capabilities to dynamically track and assess strategic readiness risks.

---

[42] The Aim Point Investment Model examines the trade-offs between modernization and readiness of Army units. A study (not available to the general public) built an optimizer to synchronize budget execution, Class IX supplies, and permanent change of station moves with units' deployment schedules.

CHAPTER 4

Conclusions and Next Steps

In this report, we sought to achieve three objectives. First, we sought to make strategic readiness conceptually easier to operationalize across department processes. There is a shared appreciation for the need to do strategic readiness well but not a shared understanding of what that means in practice. We provided a logic model that shows how the ten strategic readiness dimensions build to produce warfighting capabilities to meet strategic objectives. Chapter 3 can be read as a stand-alone chapter for guiding professional military education instruction on department processes. We also recommend that this conceptualization of strategic readiness be used to frame National Security Council Deputies Committee and Interagency Policy Committees dialogue about the impacts of proposed defense policies; guide department discussions about budget, force employment, and capability development; and inform congressional budget appropriation deliberations.

Second, we drew from risk management best practices for recommendations about how to better manage strategic readiness risk. We found that risk failures in large, conglomerate, globe-spanning institutions could often be traced back to shortfalls in five best practices: shared risk appreciation, independent oversight, transparency about cross-silo risks, engaged senior leadership, and consistent follow-through on mitigation. Based on these best practices, we recommend that the department establish a stronger independent risk oversight function, develop clear risk management guidance and governance responsibilities, and grow analytic capabilities for integrated, auditable risk assessments.

Third, we provided a way DoD can think about incorporating data analytics and AI to produce strategic readiness assessments more efficiently. Producing rigorous, verifiable risk assessments of strategic readiness trade-offs and opportunity costs is a necessary piece of risk management. One of the biggest challenges is how time- and labor-intensive such analyses can be. AI LLMs and co-pilots present an opportunity to cut back on the time and labor required to produce such analyses.

If department leadership accepts these recommendations, the question remains of how to implement them? The underlying bureaucratic challenge is determining how DoD needs to be organizationally structured to implement these practices and develop these assessment capabilities. Review of risk management best practices reveals that dedicated leadership, explicit organizational authorities, and robust analytic capabilities are essential ingredients. As we gained a better appreciation for the importance and enormity of managing strategic readiness risks, it became clear that OSD was the only entity in the department that could integrate both the national policy–level conversation and the internal DoD institutional conversation, but OSD is ill-structured to perform this fundamental responsibility. We considered the Joint Staff an essential partner to OSD in this risk management role, but the Joint Staff unable to fulfill all the roles because of its limited remit in competitive advantage dimensions, such as defense industrial base mobilization, human capital policies, modernization, and business system oversight.

Although our research scope did not allow a fulsome examination of OSD organizational structure, we recommend that OSD explore several potential options:

- *Shift emphasis within the OUSD(P&R) to readiness.* Currently, OUSD(P&R) is organizationally structured and staffed toward its personnel policy functions. The least disruptive change would be to rebalance resources within OUSD(P&R) to its readiness functions. Adding more resources to the readiness functions is also an alternative. Though this would be least disruptive to existing organizations and would perhaps face the least resistance from other OSD stakeholders, there is a danger that it will not produce sufficient change. OUSD(P&R) will still confront a bifurcation of responsibilities in decision forums about how to represent two different and, at times, mutually contradicting policy positions.
- *Spin off readiness to OUSD(Policy).* This would be more disruptive and would perhaps face resistance from OUSD(P&R) (which does not want to see its clout diminished) and from OUSD(Policy) (which may not want to take on new, unfamiliar responsibilities). The advantage of this approach is that it may be easier to integrate strategic readiness considerations in the policymaking process when the two functions reside within the same OUSD shop. Moreover, there are significant synergies between OUSD(Policy) oversight roles for plans, posture, and force development and strategic readiness, particularly focused on the middle-term time frame. A reconfigured OUSD(P&R) is perhaps the most logical focal point for strategic readiness because of the already substantial role that OUSD(Policy) has in DoD processes, such as plans, global force management, budget review, and Defense Management Action Groups. However, a criticism of this option is that it would not necessarily be independent oversight, particularly if OUSD(Policy) had strong policy preferences regarding how to prioritize strategic objectives and time frames.
- *Create an Office of the Under Secretary of Defense for Strategic Readiness.* This would be most disruptive and would likely face resistance not only from OUSD(P&R) (which would be losing much of its portfolio) but also from the other OUSD directorates—who may interpret it as an entity with an outsized evaluation voice over their policy analysis and recommendations. The advantage of this approach is that it allows the strategic readiness function to act as an integrator for all OUSD directorates and provide independent oversight across them, including OUSD(Policy), with an integrative, evaluative role for strategic readiness similar to CAPE's role in budget polices.

- *Create an Office of the Assistant to the Secretary of Defense for Strategic Readiness (OATSD[SR]) that reports directly to the DSD.* There are several Assistant Secretary of Defense–level directorates that report directly to the DSD, including Legislative Affairs. Creating a direct report OATSD(SR) would have all the advantages of creating an OUSD for strategic readiness—with equidistance from the other OSD components allowing it to be a more effective integrator and more independent overseer—but perhaps with less resistance from the other OUSD directorates. OATSD(SR) would be a more junior player but could potentially enhance its independent oversight role versus the typical policy-generating role of OUSD directorates. What matters is less whether an independent risk oversight function sits in an OUSD or OATSD but that independent oversight entity has direct access to the DSD. Being an OATSD could allow the strategic readiness function to be nimbler and more responsive to assessment questions from the DSD or with regular briefs to the DSD for trends in the department's strategic readiness profile. It would be beneficial if OATSD could be seen as more of a team player assisting the DSD and Under Secretaries of Defense in making better decisions by facilitating information-sharing about the risks, impacts, and potential negative consequences of decisions being considered across all corners of the Pentagon.

When we set out to write this report, the purpose was not to suggest an overhaul of OSD structures. The project began as an effort to develop a strategy for how DoD could better manage strategic readiness risks within the constraints of the current organizational structure with the resources that were available to it. We did not conduct a thorough review of the pros and cons of each reorganization, and we lack sufficient analysis to make a recommendation.

In this report, we have answered questions about what risk management of strategic readiness means and how to do it well. But in doing so, we have raised more questions regarding whether DoD is appropriately structured (processes, authorities, organizational roles) to effectively conduct risk management of strategic readiness. The answer may be one of the four possible changes we suggested above or a change that we have not yet conceived. Future research should pursue this line of inquiry more explicitly because one could argue that today's global threat environment has made sound risk management of strategic readiness more important than ever.

Abbreviations

AI	artificial intelligence
AR	Army Regulation
CAPE	Cost Assessment and Program Evaluation
CCMD	combatant command
CJCS	Chairman of the Joint Chiefs of Staff
CPG	Contingency Planning Guidance
CRA	Chairman's Risk Assessment
CRS	Chairman's Readiness System
DoD	U.S. Department of Defense
DoDD	Department of Defense Directive
DoDI	Department of Defense Instruction
DPG	Defense Planning Guidance
DRRS	Defense Readiness Reporting System
DSD	Deputy Secretary of Defense
ERMG	Executive Readiness Management Group
GAO	Government Accountability Office
GEF	Guidance for the Employment of the Force
GFMAP	Global Force Management Action Plan
JRAM	Joint Risk Analysis Methodology
LLM	large language model
NDS	National Defense Strategy
NSS	National Security Strategy
OASD(R)	Office of the Assistant Secretary of Defense for Readiness
OATSD(SR)	Office of the Assistant to the Secretary of Defense for Strategic Readiness
OPLAN	operational plan
OSD	Office of the Secretary of Defense
OSD(Policy)	Office of the Secretary of Defense for Policy
OUSD(P&R)	Office of the Under Secretary of Defense for Personnel and Readiness
OUSD(Policy)	Office of the Under Secretary of Defense for Policy
PPBE	planning, programming, budgeting, and execution
RDIM	Readiness Decision Impact Model
SDOB	Secretary of Defense Orders Book
SRA	Strategic Readiness Assessment

SROC	Senior Readiness Oversight Council
USD(P&R)	Under Secretary of Defense for Personnel and Readiness

References

Army Regulation 525-30, *Army Strategic and Operational Readiness*, U.S. Army, April 9, 2020.

Bernstein, Peter L., *Against the Gods: The Remarkable Story of Risk*, Wiley, 1998.

Best, Katharina Ley, Scott R. Stephenson, Susan A. Resetar, Tim Wu, Peter Schirmer, Nathaniel Edenfield, Lewis Schneider, and Emmi Yonekura, *Climate and Readiness: Quantifying Climate Effects on U.S. Joint Force Training*, RAND Corporation, RR-A2921-2, forthcoming.

Betts, Richard K., "Military Readiness: Concepts, Choices, Consequences," Brookings Institution Press, February 1, 1995.

Boito, Michael, Tim Conley, Joslyn Fleming, Alyssa Ramos, and Katherine Anania, *Expanding Operating and Support Cost Analysis for Major Programs During the DoD Acquisition Process: Legal Requirements, Current Practices, and Recommendations*, RAND Corporation, RR-2527-OSD, 2018. As of January 1, 2025:
https://www.rand.org/pubs/research_reports/RR2527.html

Bracken, Paul, Ian Bremmer, and David Gordon, eds., *Managing Strategic Surprise: Lessons from Risk Management and Risk Assessment*, Cambridge University Press, 2008.

Briggs, Gary J., *Harnessing Constructive Simulations for Reinforcement Learning*, RAND Corporation, RR-A1722-6, August 8, 2024. As of September 25, 2024:
https://www.rand.org/pubs/research_reports/RRA1722-6.html

Buehler, Kevin, Andrew Freeman, and Ron Hulme, "The Risk Revolution," McKinsey, 2008a.

Buehler, Kevin, Andrew Freeman, and Ron Hulme, "The New Arsenal of Risk Management," *Harvard Business Review*, September 2008b.

Chairman of the Joint Chiefs of Staff, "CJCS Guide to the Chairman's Readiness System," November 15, 2024.

Chairman of the Joint Chiefs of Staff Manual 3105.01B, *Joint Risk Analysis Methodology*, December 22, 2023.

CJCS—*See* Chairman of the Joint Chiefs of Staff.

Conway, Maree, *Foresight: An Introduction; A Thinking Futures Reference Guide*, Yuma Community College District, 2015.

Davis, Paul K., *Analysis to Inform Defense Planning Despite Austerity*, RAND Corporation, RR-482-OSD, March 20, 2014. As of September 23, 2024:
https://www.rand.org/pubs/research_reports/RR482.html

Davis, Paul K., *Capabilities for Joint Analysis in the Department of Defense: Rethinking Support for Strategic Analysis*, RAND Corporation, RR-1469-OSD, 2016. As of September 23, 2024:
https://www.rand.org/pubs/research_reports/RR1469.html

Davis, Paul K., Jonathan Kulick, and Michael Enger, *Implications of Modern Decision Science for Military Decision-Support Systems*, RAND Corporation, MG-360-AF, 2005. As of September 23, 2024:
https://www.rand.org/pubs/monographs/MG360.html

Department of Defense Directive 5105.79, *DoD Senior Governance Framework*, U.S. Department of Defense, November 8, 2021.

Department of Defense Directive 5124.02, *Under Secretary of Defense for Personnel and Readiness*, U.S. Department of Defense, June 23, 2008.

Department of Defense Directive 5149.2, *Subject: Senior Readiness Oversight Council (SROC)*, Office of the Director of Administration and Management, July 23, 2002.

Department of Defense Directive 7730.65, *DoD Readiness Reporting System*, U.S. Department of Defense, May 31, 2023.

Department of Defense Instruction 3000.18, *Strategic Readiness*, U.S. Department of Defense, November 30, 2023.

Department of Defense Instruction 8510.01, *Risk Management Framework for DoD Systems*, U.S. Department of Defense, July 19, 2022.

Deputy Secretary of Defense, "Implementing a Strategic Readiness Approach," memorandum, May 13, 2022, Not available to the general public.

DoD—*See* U.S. Department of Defense.

DoDD—*See* Department of Defense Directive.

DoDI—*See* Department of Defense Instruction.

DSD—*See* Deputy Secretary of Defense.

Enthoven, Alain C., and K. V. Smith, *How Much Is Enough? Shaping the Defense Program, 1961–1969*, RAND Corporation, CB-403, October 20, 2005. As of September 24, 2023: https://www.rand.org/pubs/commercial_books/CB403.html

Evrin, Volkan, "Risk Assessment and Analysis Methods: Qualitative and Quantitative," Information Systems Audit and Control Association, April 28, 2021.

Flanigan, Kevan, and Adam Regelbrugge, "Enterprise Risk Management (ERM): The Modern Approach to Managing Risks," Deloitte, September 25, 2023.

Freedberg, Sydney J., Jr., "CDAO Opens Advana to Analytics to Multiple Vendors in a Push to Scale Up," *Breaking Defense*, July 9, 2024.

GAO—*See* U.S. Government Accountability Office.

Gerring, John, *Social Science Methodology: A Unified Framework*, 2nd ed., Cambridge University Press, 2011.

Grabo, Cynthia M., *Anticipating Surprise: Analysis for Strategic Warning*, University Press of America, 2004.

Gray, Colin, *Strategy and Defence Planning: Meeting the Challenge of Uncertainty*, Oxford University Press, 2016.

Greenfield, Victoria A., Valerie L. Williams, and Elisa Eiseman, *Using Logic Models for Strategic Planning and Evaluation: Application to the National Center for Injury Prevention and Control*, RAND Corporation, TR-370-NCIPC, March 30, 2006. As of September 23, 2024: https://www.rand.org/pubs/technical_reports/TR370.html

Heuer, Richard B., *Psychology of Intelligence Analysis*, Center for the Study of Intelligence, 1999.

Hicks, Kathleen, "Creating Data Advantage," memorandum, U.S. Department of Defense, May 5, 2021.

Jackson, Kimberly, and David H. Berger, "Readiness Redefined: Now What?" *War on the Rocks*, June 12, 2023.

Kaplan, Robert S., and Anette Mikes, "JP Morgan's Loss: Bigger than 'Risk Management,'" *Harvard Business Review*, May 23, 2012a.

Kaplan, Robert S., and Anette Mikes, "Managing Risks: A New Framework," *Harvard Business Review*, June 2012b.

Larson, Eric V., *Force Planning Scenarios, 1945–2016: Their Origins and Use in Defense Strategic Planning*, RAND Corporation, RR-2173/1-A, 2019. As of January 1, 2025:
https://www.rand.org/pubs/research_reports/RR2173z1.html

Lin, Grace, "Meet Advance: How the Department of Defense Solved Its Data Interoperability Challenges," *Government Technology Insider*, April 7, 2021.

Little, Greg, Jerimiah Bennett, and Kevin Winnike, "Advana Procurement Data," U.S. Department of Defense, 2022.

Mahnken, Thomas, "Thinking About Competitive Strategies," in Thomas Mahnken, ed., *Competitive Strategies for the 21st Century: Theory, History, and Practice*, Stanford University Press, 2012.

Mazarr, Michael J., *Rethinking Risk in National Security: Lessons of the Financial Crisis for Risk Management*, RAND Corporation, CB-549, June 1, 2016. As of September 23, 2024:
https://www.rand.org/pubs/commercial_books/CB549.html

Mazarr, Michael J., *Understanding Deterrence*, RAND Corporation, PE-295-RC, April 19, 2018. As of October 22, 2024:
https://www.rand.org/pubs/perspectives/PE295.html

Mazarr, Michael J., Katharina Ley Best, Burgess Laird, Eric V. Larson, Michael E. Linick, and Dan Madden, *The U.S. Department of Defense's Planning Process: Components and Challenges*, RAND Corporation, RR-2173/2-A, 2019. As of December 28, 2024:
https://www.rand.org/pubs/research_reports/RR2173z2.html

Oates, Tim, *A Cambridge Approach to Improving Education: Using International Insights to Manage Complexity*, Cambridge University Press, 2017.

Obis, Anastasia, "DoD Preparing to Recompete for Advana," *Federal News Network*, July 16, 2024.

Pergler, Martin, "Enterprise Risk Management: What's Different in the Corporate World and Why," McKinsey, 2012.

Pergler, Martin, and Eric Lamarre, "Upgrading Your Risk Assessment for Uncertain Times," McKinsey, January 2009.

Pherson, Katherine, and Randolph Pherson, *Critical Thinking for Strategic Intelligence*, Sage, 2013.

Popper, Steven W., "Reflections: DMDU and Public Policy for Uncertain Times," in Vincent A. W. J. Marchau, Warren E. Walker, Pieter J. T. M. Bloemen, and Steven W. Popper, eds., *Decision Making Under Deep Uncertainty: From Theory to Practice*, Springer, 2018.

Savitz, Scott, Miriam Matthews, and Sarah Weilant, *Assessing Impact to Inform Decisions: A Toolkit on Measures for Policymakers*, RAND Corporation, TL-263-OSD, July 25, 2017. As of September 23, 2024:
https://www.rand.org/pubs/tools/TL263.html

Schirmer, Peter, *Computer Simulation of General and Flag Officer Management: Model Description and Results*, RAND Corporation, TR-702-OSD, August 6, 2009. As of September 25, 2024:
https://www.rand.org/pubs/technical_reports/TR702.html

Schirmer, Peter, and Jasmin Léveillé, *AI Tools for Military Readiness*, RAND Corporation, RR-A449-1, September 20, 2021. As of September 25, 2024:
https://www.rand.org/pubs/research_reports/RRA449-1.html

Schwartz, Peter, and Doug Randall, "Ahead of the Curve: Anticipating Strategic Surprise," in Francis Fukuyama, ed., *Blindside: How to Anticipate Forcing Events and Wild Cards in Global Politics*, Brookings Institution Press, 2007.

U.S. Code, Title 10, Section 117, Readiness Reporting System.

U.S. Code, Title 10, Section 136, Under Secretary of Defense for Personnel and Readiness.

U.S. Code, Title 10, Section 153, Exchange of Material and Disposal of Obsolete, Surplus, or Unclaimed Property.

U.S. Department of Defense, *DoD Dictionary of Military and Associated Terms*, March 2017.

U.S. Department of Defense, *2022 National Defense Strategy of the United States of America*, 2022.

U.S. Department of Defense, *Risk, Issue, and Opportunity: Management Guide for Defense Acquisition Programs*, September 2023.

U.S. Department of Defense 7000.14-R, *Department of Defense Financial Management Regulation*, November 2020.

U.S. Government Accountability Office, *Military Readiness: Improvement in Some Areas, but Sustainment and Other Challenges Persist*, GAO-23-106673, May 2, 2023.

Watts, Stephen, Ashley L. Rhoades, Michael E. Linick, Katharina Ley Best, Josyln Fleming, Paul W. Mayberry, John C. Jackson, and Michelle D. Ziegler, *Rethinking Strategic Readiness: A Framework for a Strategic Readiness Assessment*, RAND Corporation, RR-A2060-1, May 22, 2024. As of September 23, 2024:
https://www.rand.org/pubs/research_reports/RRA2060-1.html

Williams, Lauren C., "Pentagon Pauses Development of Its Go-To Data Analytics Tool," *Defense One*, June 10, 2024.

9 781977 414731